T0173634

AQA GCSE
Science (9–1)
Equations Practice Pack

Peter Edmunds

William Collins' dream of knowledge for all began with the publication of his first book in 1819.
A self-educated mill worker, he not only enriched millions of lives, but also founded a flourishing publishing house.
Today, staying true to this spirit, Collins books are packed with inspiration, innovation and practical expertise.
They place you at the centre of a world of possibility and give you exactly what you need to explore it.

Collins. Freedom to teach.

Published by Collins
An imprint of HarperCollins*Publishers*
The News Building
1 London Bridge Street
London
SE1 9GF

HarperCollins*Publishers*
1st Floor, Watermarque Building, Ringsend Road, Dublin 4, Ireland

Browse the complete Collins catalogue at
www.collins.co.uk

10 9 8 7 6 5 4 3

ISBN 978-0-00-845851-5

British Library Cataloguing-in-Publication Data
A catalogue record for this publication is available from the British Library.

Author: Peter Edmunds
Publisher: Katie Sergeant
Content editor: Tina Pietron
Copyeditors: David Hemsley and Jan Schubert
Proofreader: Peter Batty
Answer checkers: David Hemsley, Jan Schubert and
        Pete Robinson CSciTeach
Cover designer: Ken Vail Graphic Design Ltd
Cover image: Roman Sigaev/Shutterstock
Internal designer and typesetter: Ken Vail Graphic
        Design Ltd
Production controller: Katharine Willard
Printed and Bound in the UK using 100% Renewable
Electricity at CPI Group (UK) Ltd

This book is produced from independently certified FSC™ paper to ensure responsible forest management.

For more information visit:
www.harpercollins.co.uk/green

Acknowledgements
The publishers gratefully acknowledge the permission granted to reproduce the copyright material in this book. Every effort has been made to trace copyright holders and to obtain their permission for the use of copyright material. The publishers will gladly receive any information enabling them to rectify any error or omission at the first opportunity.

MIX
Paper from
responsible sources
FSC™ C007454

Contents

> **HT: Higher tier only**

How to use this book

Practice makes perfect.

Practice makes permanent.

This book is built around the concept of deliberate practice. The more we practise something, the better we get at it.

Students often get revision wrong. They read the textbook. They make colourful notes. However, if they don't practise recalling their knowledge and using their skills, then it won't be worthwhile.

If we wanted to get good at playing the guitar, we wouldn't read about playing the guitar. We would *practise* playing the guitar. If we wanted to be a professional footballer, we wouldn't read about how to become a professional footballer. We would *practise* playing football. It's the same with science. If we want to become good at answering scientific questions, then we need to *practise* answering scientific questions.

This book is built around this premise and offers a vast amount of practice in the mathematical areas of science.

The book is split into five sections:

1. Mathematics in science

2. Biology

3. Chemistry

4. Physics (paper 1)

5. Physics (paper 2)

Materials and links

This book contains sheets designed for student self-quizzing of the equations. These are at the beginning of each of the biology, chemistry and physics sections. Students should be encouraged to use these frequently so that memorisation of the equations is easier.

You will find indications in the contents and on the relevant equation double-page spread regarding the relevance to GCSE Combined Science Trilogy Foundation and Higher tiers.

All the introductory material and questions and answers can be downloaded in Microsoft and Google editable format at www.collins.co.uk/ScienceEquationsPractice/download

Differentiation

Each section is made up of worksheets designed for student practice. They are differentiated into three (or four) sections; basic, medium and hard (with the addition of a mixed practice section in certain sheets).

The basic questions are designed to be more accessible. They could be practising using an equation, but without additional complexities like unit conversions or rearranging of the equation. If done correctly, the basic questions should scaffold students into the medium difficulty questions.

The reasons why the medium questions are harder are made clear in each worksheet. It's worth emphasising this reason to students. Hints for some medium questions are in dashed boxes. If students find it difficult, it's important that they realise *why* the section is more difficult and practise that area.

Harder questions usually involve more than one complication and can get quite difficult. They could involve rearranging *and* unit conversion, for example (without the hints).

Finally, certain worksheets contain a section for mixed practice. The more frequently a student retrieves prior taught knowledge, the better they can recall it in the future. To enhance retrieval, these sections rely on previous sections and could, for example, include a multi-step calculation that may include using more than one equation.

Equations in the exam

Black boxes give information about whether the equation will be given in the exam or if students need to remember it. Unless stated otherwise, all answers should be given to two significant figures.

The book is tailored to the AQA specification, and follows this in order (although, of course, you might want to cover in a different order).

Indeed, mathematical skills make up a substantial amount of the marks in all AQA GCSE science examinations. Biology, chemistry and physics examinations respectively have 10%, 20% and 30% of the total marks awarded. Combined science has an average of 20%, in the same ratio split for each subject.

In physics, there are a total of 23 equations that students need to be able to remember and use (21 in the combined science physics papers). There are an additional 12 equations that students need to be able to select from an equation sheet and use in the exam (7 in the combined science physics papers).

In low-demand questions (in foundation tier exams) on physics papers, students will be given the equation to use (regardless of whether it is one that students are expected to remember). Medium-demand questions (as in the latter parts of a foundation tier paper or in the earlier parts of a higher tier paper) will prompt students to write the equation that is necessary (before the calculation itself). This prompt is phased out in high-demand questions (in the latter parts of a higher tier paper).

In AQA GCSE science examinations, the prompt "write down any equations you use" will be used for any calculation questions that involve the use of more than one equation.

 © HarperCollins*Publishers* Limited 2021

How to answer equation questions

To maximise chances of getting marks, it's vital that you show clear working at all times while undertaking calculation questions. To do this, it is recommended you follow a method like the one below:

1. Write down the equation

2. Substitute the variables into the equation (doing any unit conversions if needed)

3. Rearrange (if needed)

4. Calculate the answer

5. Write down the units

A sample question is given below. This one requires use of the equation $F = ma$ (*force = mass × acceleration*):

A radio-controlled car has a mass of 0.4 kg, and an acceleration of 2 m/s². Find the resultant force that is acting on the car. [3 marks]

Write down the equation:	$F = ma$	
Substitute the variables into the equation:	$F = 0.4 × 2$	[1 mark]
Calculate the answer:	$F = 0.8$	[1 mark]
Remember to add the units:	newtons (N)	[1 mark]

The following shows a slightly more complicated question which requires some rearrangement of the equation. Here the equation $\rho = \frac{m}{V}$ $\left(density = \frac{mass}{volume}\right)$ is required:

A bath is filled with 0.15 m³ of water. Calculate the mass of the water in the bath. Water has a density of 1000 kg/m³. [3 marks]

Write down the equation:	$\rho = \frac{m}{V}$	
Substitute the variables into the equation:	$1000 = \frac{m}{0.15}$	[1 mark]
Rearrange the equation to make m the subject. Note how the right-hand side of the equation is "$\frac{m}{0.15}$". To get m by itself, you can multiply both sides by 0.15:	$1000 × 0.15 = m$	
Calculate the answer:	$m = 150$	[1 mark]
Remember to add the units:	kilograms (kg)	[1 mark]

This final sample question involves both rearranging and a unit conversion, together with the equation $P = VI$ (*power = potential difference × current*):

A lamp has a current of 100 mA flowing through it and a power of 0.6 W. Find the potential difference across the lamp. [4 marks]

Write down the equation:	$P = VI$	
Convert the current into amps:	100 mA = 0.1 A	[1 mark]
Substitute the variables into the equation:	$0.6 = V \times 0.1$	[1 mark]
Rearrange the equation to make V the subject. Note how the right-hand side of the equation is "0.1 × V". To get V by itself, you divide both sides by 0.1:	$\frac{0.6}{0.1} = V$	
Calculate the answer:	$V = 6$	[1 mark]
Remember to add the units:	volts (V)	[1 mark]

After you have calculated the answer, it's important to have a look at the order of magnitude of the answer. If you had forgotten the unit conversion for the last question, you would have obtained an answer of 0.006 V. That's a very small potential difference and should have given you cause for concern.

Similarly, if the mass of a bus is calculated to be something like 0.5 kg then that's a sure sign that something has gone wrong. Checking the feasibility of an answer can be really important in noticing mistakes!

Significant figures and standard form

Significant figures are often used in science. Every measurement has an error associated with it. The number of significant figures gives an indication of the precision of a measurement or number.

Scientists also sometimes need to use very small or large numbers. To make writing and using these more convenient, use standard form.

Basic questions: Basic standard form and significant figures

When writing a number in standard form there are two parts; a number between 1 and 10, and the power of ten. For example, you can write the speed of light (300 000 000 m/s) as 3×10^8 m/s.

Q1. Write the following numbers in standard form:

    a) 5000   **b)** 24 000   **c)** 170   **d)** 150 000   [1 mark each]

    a) ....

    b) ....

    c) ....

    d) ....

When writing numbers in significant figures, the "decider digit" tells you what to round to. For example, say you need to round the time 23.42 s to three significant figures. Here, the decider digit is the fourth digit (i.e. the number 2). As this is less than five, you round down to 23.4 s. If it was above five, then you would round up.

Q2. Write the following numbers to two significant figures:

    a) 432 000   **b)** 324   **c)** 93 700   **d)** 432 452   [1 mark each]

    a) ....

    b) ....

    c) ....

    d) ....

Medium questions: Slightly harder standard form and significant figures

You can also write small numbers in standard form, by writing a negative power of ten (which is the same as dividing by that power of ten). For example, you can write the distance 0.00045 m as 4.5×10^{-4} m.

Q3. Write the following numbers in standard form:

    a) 0.023   **b)** 0.0032   **c)** 0.56   **d)** 0.00873   [1 mark each]

    a) ....

    b) ....

    c) ....

    d) ....

Q4. Complete the table and write each number to the required number of significant figures.   [1 mark each]

Number	To three significant figures	To two significant figures	To one significant figure
4324			
8431			
0.274			
4.308			
0.00239			

Hard questions: Significant figures and standard form combined

Q5. Write the number 275 000 000 in standard form and to two significant figures.   [2 marks]

....

....

Q6. Write the number 0.000837 in standard form and to two significant figures.   [2 marks]

....

....

Q7. Write the number 8739 in standard form and to three significant figures.   [2 marks]

....

....

Q8. Write the number 0.0189 in standard form and to one significant figure.   [2 marks]

....

....

Q9. Write the number 439 200 in standard form and to two significant figures.   [2 marks]

....

....

Q10. Write the number 0.242 in standard form and to three significant figures.   [2 marks]

....

....

[Total marks   / 39]

Rearranging equations

On the previous pages, you learned the general strategy for answering basic equation-based questions. However, the equation that you need to use will not always be in the correct format. Sometimes, you need to *rearrange* an equation so that it can be used correctly. These pages will go over how to do this.

Basic questions: Basic rearranging of equations like $P = VI$

Take the equation $P = VI$. Here, P is the symbol for the power of an electronic device (with units of watts (W)). I is the symbol for current through the device (with units of amps (A)), and V is the potential difference across the device (with units of volts (V)).

Model example: A lamp is connected to a 6 V battery and has a power of 24 W. Calculate the current through the lamp. [3 marks]

Write down the equation:	$P = VI$	
Substitute the variables into the equation:	$24 = 6 \times I$	[1 mark]
Rearrange the equation to make I the subject. Note how the right-hand side of the equation is "6 × I". To get I by itself, you can divide both sides by 6:	$\frac{24}{6} = I$	
Calculate the answer:	$I = 4$	[1 mark]
Remember to add the units:	amps (A)	[1 mark]

Q1. An LED is connected to a 12 V power supply. Calculate the current for each power:

a) 6 W   **b)** 24 W   **c)** 18 W   **d)** 90 W   [3 marks each]

a) ....

b) ....

c) ....

d) ....

Q2. A motor has a power of 60 W, and a current of 1.2 A flowing through it. Calculate the potential difference across the motor. [3 marks]

....

Medium questions: Rearranging of equations like $P = \frac{W}{t}$

Take the equation $P = \frac{W}{t}$. The power is again denoted by P, and W is the work done (with units of joules (J)) and t is the time (with units of seconds (s)).

Model example: An electric motor has a power of 2000 W and does 40 000 J of work. Calculate the time for which the motor is on. [3 marks]

Write down the equation:	$P = \frac{W}{t}$	
Substitute the variables into the equation:	$2000 = \frac{40\,000}{t}$	[1 mark]

Rearrange the equation to make t the subject. Firstly, multiply both sides by t, then divide both sides by 2000:	$2000 \times t = 40\ 000$ $$t = \frac{40\ 000}{2000}$$
Calculate the answer:	$t = 20$ [1 mark]
Remember to add the units:	seconds (s) [1 mark]

Q3. A kettle has a power of 2500 W. Calculate the time taken to do the following amounts of work:

a) 50 000 J   b) 1 000 000 J   c) 75 000 J   d) 600 000 J   [3 marks each]

a) ....

b) ....

c) ....

d) ....

Hard questions: More complex rearranging of equations like $E_k = \frac{1}{2}mv^2$

The equation $E_k = \frac{1}{2}mv^2$ gives the amount of kinetic energy a moving object has (with units of joules (J)). The symbol m gives the mass (with units of kg) and v is the velocity (with units of m/s).

Model example: A van has a kinetic energy of 150 000 J and a mass of 3000 kg. Calculate the velocity of the van.   [3 marks]

Write down the equation:	$E_k = \frac{1}{2}mv^2$
Substitute the variables into the equation, then simplify the right-hand side of the equation:	$150\ 000 = 0.5 \times 3000 \times v^2$ $150\ 000 = 1500 \times v^2$ [1 mark]
Rearrange the equation to make v^2 the subject; divide both sides by 1500. Then simplify the left-hand side:	$\frac{150\ 000}{1500} = v^2$ $100 = v^2$
To calculate the answer, take the square root of both sides of the equation:	$10 = v$ [1 mark]
Remember to add the units:	meters per second (m/s) [1 mark]

Q4. A car has a kinetic energy of 120 000 J. Calculate the velocity of the car for the following masses:

a) 800 kg   b) 1200 kg   c) 900 kg   d) 1600 kg   [3 marks each]

a) ....

b) ....

c) ....

d) ....

[Total marks   / 39]

Mean and uncertainties

In scientific experiments, every measurement that you take has an uncertainty associated with it. Every time you make a measurement, it will be slightly different. This is due to random errors. To reduce the effect of random errors, you can repeat a measurement multiple times and take a mean value.

Basic questions: Calculations of mean

To calculate the mean you add together the individual values and divide by the total number of values:

$$Mean = \frac{sum\ of\ values}{number\ of\ values}$$

Model example: An ammeter measures how much current is flowing through a circuit.
Four measurements were obtained: [1 mark]

1.2 A   **1.1 A**   **1.2 A**   **1.3 A**

Calculate the mean value of the current in the circuit.

You can answer this question in the following way:

Calculate the sum of values: 1.2 + 1.1 + 1.2 + 1.3 = 4.8 A

Divide this by the number of values: 4.8 ÷ 4 = 1.2 A   [1 mark]

Q1.   Calculate the mean of the following potential differences: [1 mark]

3.2 V   **2.9 V**   **3.1 V**   **3.0 V**   **3.3 V**

....

Q2.   The number of paperclips that an electromagnet can pick up is measured. From the results below, calculate the mean number of paperclips that an electromagnet can pick up.   [1 mark]

5   **6**   **7**   **6**

....

Medium questions: Identifying anomalies and calculating uncertainties

Sometimes an experiment gives a result that doesn't fit in with the other results. This is called an anomaly. You disregard the anomaly from the results when calculating (for example) a mean.

Q3.   Identify the anomaly and calculate the mean of the following currents to three significant figures:

[2 marks]

12.9 A   **11.5 A**   **12.8 A**   **13.1 A**   **13.2 A**

....

Q4.   Identify the anomaly and calculate the mean of the following resistances to three significant figures:

[2 marks]

50.2 Ω   **50.5 Ω**   **60.5 Ω**   **49.9 Ω**   **49.8 Ω**

....

You can also use multiple measurements to obtain the *uncertainty*. To do this, you can use the following equation: $uncertainty = \frac{range}{2}$

where the range = maximum value – minimum value

Model example: Take the same measurements of the current again:

| **1.2 A** | **1.1 A** | **1.2 A** | **1.3 A** |

You already know that the mean is 1.2 A. Now calculate the uncertainty. [2 marks]

Calculate the range: 1.3 – 1.1 = 0.2 A [1 mark]

Divide this by 2 to get the uncertainty: $\frac{0.2}{2} = 0.1$ A [1 mark]

Write the answer as mean ± uncertainty: 1.2 ± 0.1 A

Q5. Find the mean and the uncertainty of the following temperatures. Give your answer to three significant figures: [3 marks]

| **37.5 °C** | **36.8 °C** | **37.2 °C** | **37.0 °C** | **36.5 °C** |

....

....

Q6. Find the mean and the uncertainty of the following masses: [3 marks]

| **0.5 kg** | **0.6 kg** | **0.7 kg** | **0.6 kg** | **0.6 kg** |

....

....

Hard question: Rearranging and calculating uncertainties

Q7. The equation that links potential difference (V), current (I) and resistance (R) is $V = IR$. Potential differences across a resistor and current flowing through a resistor were measured. The results are shown in the table. Calculate the mean resistance and uncertainty. Space is given in the table to write the resistance for each set of readings. Calculate all answers to two decimal places. [6 marks]

Potential difference (V)	Current (A)	Resistance (Ω)
1.0	1.2	
2.0	2.3	
3.0	2.9	
4.0	4.8	

....

....

[Total marks / 18]

Percentages and percentage changes

One "per cent" means that you have one part per hundred.

Percentages are a useful tool to compare the size of different quantities; you need to be able to calculate percentages and percentage changes of quantities.

$Percentage\ change = \frac{difference}{original} \times 100$

Basic questions: Calculating percentages

Model example: Find 60 as a percentage of 75.　　　　　　　　　　　　　　　　　　[1 mark]

Divide 60 by 75:　　　　　　　　　　　　　　　$\frac{60}{75} = 0.8$

Multiply this by 100:　　　　　　　　　　　　$0.8 \times 100 = 80\%$　　　　　　　　　　[1 mark]

Q1. Find 50 as a percentage of:

　　a) 80　　　　　　**b)** 200　　　　　**c)** 1250　　　　　**d)** 50　　　　　[1 mark each]

　　a) ....

　　b) ....

　　c) ....

　　d) ....

Q2. Find 40 as a percentage of:

　　a) 80　　　　　　**b)** 4000　　　　**c)** 200 000　　　**d)** 500　　　　[1 mark each]

　　a) ....

　　b) ....

　　c) ....

　　d) ....

Medium questions: Calculating percentage changes

Model example: The speed of a car accelerates from 50 to 60 m/s. Find the percentage change in the speed of the car.　　　　　　　　　　　　　　　　　　　　　　　　　　[2 marks]

Calculate the difference in speed:　　　　　　60 – 50 = 10 m/s

Divide the difference in speed by the original speed:　　　　　　　　　　$\frac{10}{50} = 0.2$　　　　　　　　　　[1 mark]

To calculate the percentage change, multiply this by 100:　　　　　$0.2 \times 100 = 20\%$　　　　　　　　[1 mark]

Q3. A car is initially travelling at 30 m/s. Calculate the percentage change in speed if it accelerates to:

a) 60 m/s   **b)** 45 m/s   **c)** 90 m/s   **d)** 36 m/s   [2 marks each]

a) ....

b) ....

c) ....

d) ....

Hard questions: Harder, word-based questions

Q4. As a car accelerates, its chemical store of energy from petrol decreases by 20 000 J. The car's kinetic store of energy increases by 14 000 J. Calculate the energy transfer efficiency as a percentage.   [1 mark]

....

....

Q5. An iron bolt has an initial mass of 40 g. It is left outside in the rain and oxidises. The mass increases to 48 g. Find the percentage increase in the mass of the bolt.   [2 marks]

....

....

Q6. A sample of 2.5 g copper carbonate is heated and undergoes thermal decomposition. The final mass is 2.0 g. Find the percentage decrease in the mass of the sample.   [2 marks]

....

....

Q7. An experiment is being undertaken to investigate osmosis in potato cells. A disc of potato initially has a mass of 4.0 g. After it has been left in water, the mass increases to 4.2 g. Find the percentage increase in the mass of the potato disc.   [2 marks]

....

....

Q8. A sprinter initially has a speed of 8 m/s and accelerates to a speed of 10 m/s. Find the percentage increase in the speed of the sprinter.   [2 marks]

....

....

[Total marks   / 25]

Gradients

Gradients are often used to show the "rate" at which something is happening. For example, the gradient of a displacement–time graph gives the velocity. The steeper the gradient of a displacement–time graph, the greater the velocity. You can calculate the gradient of a graph by using the following equation:

$$gradient = \frac{change\ in\ y}{change\ in\ x} = \frac{\Delta y}{\Delta x}$$

Basic questions: Calculating gradients using given Δy and Δx

Q1. Find the gradient for each of the sets of Δy and Δx:

    a) $\Delta y = 100$, $\Delta x = 5$ **b)** $\Delta y = 15$, $\Delta x = 2.5$ **c)** $\Delta y = -2.5$, $\Delta x = 0.5$ **d)** $\Delta y = 0.8$, $\Delta x = 4$ [1 mark each]

    a) ....

    b) ....

    c) ....

    d) ....

Medium questions: Calculating gradients from linear graphs

Q2. Find the gradient for each of the graphs: [2 marks each]

a)

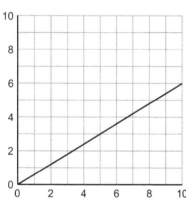

b)

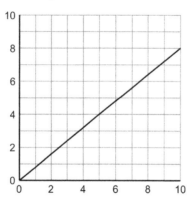

c)

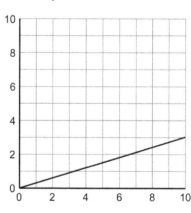

d)

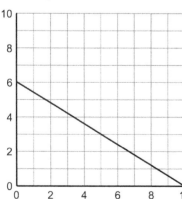

e)

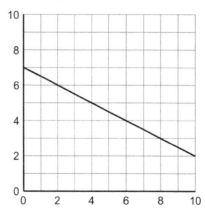

f)

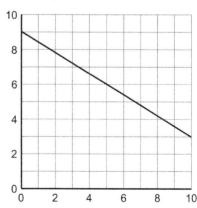

a) .... / 25]

b) ....

c) ....

d) ....

e) ....

f) ....

Hard questions: Calculating gradients using tangents on non-linear graphs

Sometimes you might need to calculate the gradient of a curve. To do this, you need to draw a *tangent* to the curve. This tangent will be a straight line and will follow the direction of the curve at that point. You can use this straight line to calculate a gradient in the same way as before.

Q3. Find the gradient of the graph at:

a) $x = 0$ **b)** $x = 6$ **c)** $x = 20$ [3 marks each]

a) ....

....

b) ....

....

c) ....

....

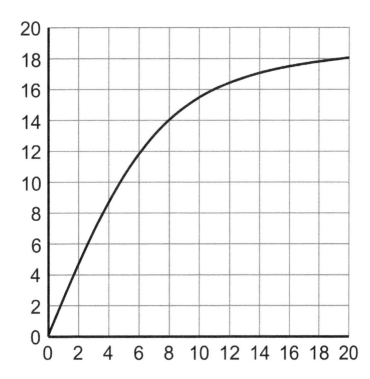

[Total marks   / 25]

Biology summary sheet

In the biology examinations, 10% of the overall marks are awarded for mathematical skills. This includes not only use of the equations on the following page, but also the general skills that you saw previously. Examples of this include calculation of means, percentages and gradients; as well as rearrangement of equations.

The seven topics in biology are:

1. Cell biology

2. Organisation

3. Infection and response

4. Bioenergetics

5. Homeostasis and response

6. Inheritance, variation and evolution

7. Ecology.

The equations on the following page are split across these topics and need to be remembered. Cover up the right hand side and regularly test yourself to check that you can remember the equations.

Note that there is some overlap in calculation requirements in biology, chemistry and physics. In particular, surface area to volume ratio and rate of change both appear within the chemistry section. Magnification can also be found within the physics section.

The equations in **bold** are biology only.

Magnification	$magnification = \dfrac{size\ of\ image}{size\ of\ real\ object}$
Number of bacteria	$number\ of\ bacteria = number\ of\ bacteria\ at\ start \times 2^{\,number\ of\ divisions}$
Cross-sectional area of a circular colony	$cross\text{-}sectional\ area = \pi r^2$
Surface area to volume ratio	$surface\ area \div volume$
Rate of change	$rate\ of\ change = \dfrac{change\ in\ value}{change\ in\ time}$
Efficiency of biomass transfer	$efficiency\ of\ biomass\ transfer = \dfrac{biomass\ of\ higher\ trophic\ level}{biomass\ of\ lower\ trophic\ level} \times 100$
Light intensity	$light\ intensity \propto \dfrac{1}{distance^2}$

Magnification (combined science)

Microscopes allow us to see objects that are too small to see with our eyes alone.

An electron microscope uses electrons instead of light (like a light microscope) to form an image.

The equation to calculate magnification is: $magnification = \frac{size\ of\ image}{size\ of\ real\ object}$

You need to remember this equation!

Basic questions: No rearranging or unit conversion needed.

Model example: An onion cell is imaged under a light microscope. The image formed is of length 5 mm, but the real length of the onion cell is 0.25 mm. Calculate the magnification of the microscope.   [2 marks]

Write down the equation: $magnification = \frac{size\ of\ image}{size\ of\ real\ object}$

Substitute the variables into the equation: $magnification = \frac{5}{0.25}$ [1 mark]

Calculate answer: $magnification = \frac{5}{0.25} = 20$ [1 mark]

Q1. A specimen is imaged under a light microscope. The size of the image is 8 mm. Calculate the magnification for each real size:

   a) 0.5 mm   **b)** 0.16 mm   **c)** 0.2 mm   **d)** 0.1 mm   [2 marks each]

   a) ....

   b) ....

   c) ....

   d) ....

Q2. A specimen of length 0.2 mm is imaged under a light microscope. Calculate the magnification for each image length:

   a) 3 mm   **b)** 15 mm   **c)** 9 mm   **d)** 1.6 mm   [2 marks each]

   a) ....

   b) ....

   c) ....

   d) ....

Medium questions: Unit conversion needed (hints in boxes).

Q3. A cell is imaged under a light microscope and has a length of 50 μm. Calculate the magnification for each image length:

μm → mm ÷ 1000

   a) 10 mm   **b)** 15 mm   **c)** 2 mm   **d)** 1.5 mm   [3 marks each]

   a) ....

   b) ....

c) ....

d) ....

Q4. A textbook has an image of some onion cells. The onion cells have a real length

cm → mm × 10

of 0.25 mm. Calculate the magnification for each image size:

a) 2 cm   **b)** 8 cm   **c)** 5 cm   **d)** 6 cm   [3 marks each]

a) ....

b) ....

c) ....

d) ....

Hard questions: Rearranging and unit conversion needed.

Q5. A red blood cell has a diameter of 8 μm and is imaged under a microscope that has a magnification of 150. Calculate the diameter of the image in mm.   [3 marks]

....

....

Q6. In a diagram, the nucleus of a cell has a diameter of 1.2 mm. The image is drawn with a magnification of 200. Calculate the actual diameter of the nucleus in μm.   [3 marks]

....

....

Q7. A ribosome of length 20 nm is imaged under an electron microscope that has a magnification of 2 000 000. Calculate the length of the image formed in mm.   [3 marks]

....

....

Q8. Some palisade cells are imaged under a light microscope that has a magnification of 50. Twelve palisade cells have a total length of 240 μm. Calculate the length of the image of one palisade cell in mm.   [4 marks]

....

....

[Total marks   / 53]

Culturing microorganisms (biology only)

Given sufficient nutrients and a suitable temperature, bacteria can reproduce and double in number as often as every 20 minutes. You can calculate the number of bacteria at the end of the growth period by using the equation:

You need to remember these equations!

number of bacteria = number of bacteria at start × 2 *number of divisions*

If the bacteria are grown as a colony on an agar gel plate then the cross-sectional area of the colony can be calculated using the equation: *cross-sectional area = πr^2*.

Basic questions: Cross-sectional areas of colonies.

Model example: A circular colony of bacteria is on a Petri dish, with a radius of 2 cm. Calculate the cross-sectional area of the colony.   [3 marks]

Write down the equation:	*cross-sectional area = πr^2*
Substitute the variables into the equation:	*cross-sectional area = $\pi \times 2^2$* [1 mark]
Calculate answer:	*cross-sectional area = $\pi \times 2^2 = 13$* [1 mark]
Write units:	centimetres squared (cm^2) [1 mark]

Q1. A circular colony of bacteria is on a Petri dish. Calculate the cross-sectional area of the colony for each radius:

    a) 1.5 cm   **b)** 5 cm   **c)** 4 cm   **d)** 7 cm   [3 marks each]

    a) ....

    b) ....

    c) ....

    d) ....

Medium questions: Calculating numbers of bacteria.

Model example: A bacterial colony has a mean division time of 20 minutes. If there are initially 100 bacteria, calculate the number of bacteria present after 80 minutes.   [3 marks]

Calculate number of divisions	$80 \div 20 = 4$ divisions [1 mark]
Write down the equation:	*number of bacteria = number of bacteria at start* × 2 *number of divisions*
Substitute the variables into the equation:	*number of bacteria = 100×2^4* [1 mark]
Calculate answer:	*number of bacteria = $100 \times 2^4 = 1600$* [1 mark]

Q2. A bacterial colony has a mean division time of 30 minutes. If there are initially 50 bacteria, calculate the number of bacteria present after each of the times:

a) 30 minutes   **b)** 180 minutes   **c)** 120 minutes   **d)** 90 minutes   [3 marks each]

a) ....

b) ....

c) ....

d) ....

Q3. There are initially 20 bacteria in a colony. For each of the mean division times, calculate the number of bacteria present after 120 minutes:

a) 30 minutes   **b)** 60 minutes   **c)** 40 minutes   **d)** 20 minutes   [3 marks each]

a) ....

b) ....

c) ....

d) ....

Hard questions: Rearranging and unit conversion needed.

Q4. A Petri dish contains a colony of bacteria in a circle of cross-sectional area 6 cm^2. Calculate the radius of the colony in mm. [4 marks]

....

....

Q5. There are initially 2500 bacteria in a colony. The bacteria have a mean division time of 20 minutes. Calculate how many bacteria there will be after a time of 3 hours. Give your answer in standard form. [5 marks]

....

....

Q6. A doctor is checking whether a colony of bacteria is showing resistance to an antibiotic. He places a disc soaked in the antibiotic inside a Petri dish. No bacteria grew in a circle of cross-sectional area of 3.5 cm^2. Calculate the radius of the clear area in mm. [4 marks]

....

....

[Total marks   / 49]

Surface area to volume ratio (combined science)

Some organisms and single cells have a high surface area to volume ratio and are able to exchange substances directly with the environment through diffusion. Most multicellular organisms have a much lower ratio and therefore must have a dedicated exchange system. These exchange systems are adapted to increase the rate of diffusion by having a large surface area.

For example, the small intestine is long and folded, with villi and microvilli that give it a total surface area of approximately 30m^2.

The equation for surface area to volume ratio is: *surface area ÷ volume*

> **You need to remember this equation!**

Basic questions: No rearranging or unit conversion needed.

Model example: A cell has a surface area of 20 µm^2 and a volume of 5 µm^3. Calculate the surface area to volume ratio for the cell. [1 mark]

Write down equation: *surface area ÷ volume*

Substitute variables into equation: *surface area ÷ volume* = 20 ÷ 5 = 4

Write as a ratio: *surface area : volume ratio* = 4 : 1 [1 mark]

Q1. A cell has a surface area of 16 µm^2. Calculate the surface area to volume ratio for each volume:

a) 5 µm^3 b) 3.2 µm^3 c) 4 µm^3 d) 2.5 µm^3 [1 mark each]

a) ....

b) ....

c) ....

d) ....

Medium questions: Need to calculate surface area and volume individually.

Model example: Calculate the surface area to volume ratio for a cube of side length 0.5 cm. [3 marks]

Calculate the surface area of the cube: each side = 0.5 × 0.5 = 0.25 cm^2 [1 mark]

Six sides overall, so total surface area
= 6 × 0.25 = 1.5 cm^2

Calculate the volume of the cube: *volume* = 0.5 × 0.5 × 0.5 = 0.125 cm^3 [1 mark]

Calculate the surface area to volume ratio of the cube: *surface area ÷ volume* = 1.5 ÷ 0.125 = 12   [1 mark]

surface area : volume ratio = 12 : 1

Q2. Calculate the surface area to volume ratio of a cube of side length:

a) 1 cm   b) 2 cm   c) 3 cm   d) 4 cm   [3 marks each]

a) ....

....

b) ....

....

c) ....

....

d) ....

....

Q3. What happens to the surface area to volume ratio as the size of the cube increases? [1 mark]

....

Hard questions: Rearranging needed.

Q4. A red blood cell has a surface area of 136 μm^2 and a surface area to volume ratio of 1.5 : 1. Calculate the volume of a red blood cell. [3 marks]

....

....

Q5. A cell has a volume of 50 μm^3 and a surface area to volume ratio of 3 : 1. Calculate the surface area of the cell. [3 marks]

....

....

Mixed practice: Requires unit conversion and more than one equation.

Q6. The image of an amoeba is 10.5 mm long under a microscope of ×15 magnification.

a) Calculate the size of the amoeba in mm. [2 marks]

....

....

b) The microscope is now changed to a magnification of ×25. Calculate how long the image of the amoeba would appear under this magnification. [3 marks]

....

....

c) Assume that the amoeba is a cube of side length equal to what you calculated in part a). Calculate the surface area to volume ratio of the amoeba. [3 marks]

....

....

[Total marks   / 31]

Rates (combined science)

You can calculate rates for many different biological phenomena. For example; breathing rates, heart rate, rate of blood flow and rate of photosynthesis.

Generally, the following equation can be used to calculate a rate:

$$rate\ of\ change = \frac{change\ in\ value}{change\ in\ time}$$

> **You need to remember this equation!**

Basic questions: No rearranging or unit conversion needed.

Model example: An athlete takes 180 breaths in a time of 4 minutes. Calculate the athlete's average breathing rate in breaths per minute. [2 marks]

Write down the equation:
$$breaths\ per\ minute = \frac{number\ of\ breaths}{number\ of\ minutes}$$

Substitute the variables into the equation:
$$breaths\ per\ minute = \frac{180}{4}$$
[1 mark]

Calculate answer:
$$\frac{180}{4} = 45\ breaths\ per\ minute$$
[1 mark]

Q1. Calculate the average breathing rate of a cat if the cat takes 280 breaths in each of the times. Give your answer in breaths per minute.

a) 7 minutes   b) 4 minutes   c) 20 minutes   d) 15 minutes   [2 marks each]

a) ....

b) ....

c) ....

d) ....

Model example: An artery has 150 ml of blood flow through it in a time of 5 minutes. Calculate the rate of flow of blood in millilitres per minute. [2 marks]

Write down the equation:
$$rate\ of\ blood\ flow = \frac{volume\ of\ blood}{number\ of\ minutes}$$

Substitute the variables into the equation:
$$rate\ of\ blood\ flow = \frac{150}{5}$$
[1 mark]

Calculate answer:
$$\frac{150}{5} = 30\ ml/min$$
[1 mark]

Q2. A vein has 60 ml of blood flow through it. Calculate the rate of flow of blood (in ml/min) for each time:

a) 12 minutes   b) 15 minutes   c) 20 minutes   d) 8 minutes   [2 marks each]

a) ....

b) ....

c) ....

d) ....

Medium questions: Rearranging or unit conversion needed (hint in box):

Q3. During a football match, a footballer has a heart rate of 120 beats per minute. Calculate how many times the footballer's heart has beat for each of the times:

a) 3 minutes   **b)** 5 minutes   **c)** 15 minutes   **d)** 7 minutes   [2 marks each]

a) ....

b) ....

c) ....

d) ....

Q4. A dog takes 2000 breaths. Calculate the average breathing rate (in breaths per minute) of the dog for each of the times:

hours → mins × 60

a) 1 hour   **b)** 1 hour 40 mins   **c)** 2 hours 20 mins   **d)** 3 hours   [3 marks each]

a) ....

b) ....

c) ....

d) ....

Hard questions: The inverse square law and the rate of photosynthesis. **HT**

Light intensity is inversely proportional to the distance between the light source and the object squared:

$$light\ intensity \propto \frac{1}{distance^2}$$

Model example: At a distance of 10 cm from a lamp, a plant gives off 20 bubbles of oxygen per minute. Calculate how many bubbles of oxygen per minute the plant would give off at a distance of 20 cm.   [2 marks]

Write down the equation:	$light\ intensity \propto \frac{1}{distance^2}$
Substitute the variables into the equation:	$light\ intensity \propto \frac{1}{2^2}$
Calculate light intensity at new distance:	Light intensity is $\frac{1}{4}$ of the original intensity. [1 mark]
Number of bubbles of oxygen per minute also decreases by this factor:	$\frac{1}{4} \times 20 = 5$ bubbles of oxygen per minute [1 mark]

Q5. At a distance of 5 cm from a lamp, a plant gives off 36 bubbles of oxygen per minute. Calculate how many bubbles of oxygen the plant would give off for each of the distances:

a) 10 cm   **b)** 15 cm   [2 marks each]

a) ....

b) ....

[Total marks   / 40]

Genetic diagrams and probabilities (combined science)

A genetic diagram shows us the possible genotypes of offspring from given parents. Upper case letters represent dominant alleles, while lower case letters represent recessive alleles.

You need to remember this!

A dominant allele is always expressed if present, while two copies of a recessive allele must be present for the recessive allele to be expressed. Knowing the genotype allows us to make predictions for the phenotype.

An organism that carries two identical alleles can be described as homozygous, whereas an organism that carries two different alleles of the same gene is heterozygous.

Basic questions: Punnet squares given.

Model example: The dominant allele for brown hair is represented by B, while b represents the recessive allele for blonde hair. Two parents have the genotype Bb and the Punnett square represent their potential offspring. Calculate the probability (as a fraction) the child will have blonde hair.

	B	b
B	BB	Bb
b	Bb	bb

[2 marks]

Identify the blonde genotype: bb [1 mark]

Calculate the probability: Four potential genotypes. Only one is bb. [1 mark]

$\frac{1}{4}$ probability the child will have blonde hair.

Q1. The dominant allele for brown hair is represented by B, while b represents the recessive allele for blonde hair. Use each Punnett square diagram to calculate the probability the child will have brown hair:

a)
	B	b
b	Bb	bb
b	Bb	bb

b)
	B	b
B	BB	Bb
B	BB	Bb

c)
	b	b
b	bb	bb
b	bb	bb

d)
	B	b
B	BB	Bb
b	Bb	bb

[2 marks each]

a) ....

b) ....

c) ....

d) ....

Medium questions: Genetic diagram given, calculation of percentages needed.

Model example: The dominant allele for a plant being tall is represented by T, while t represents the recessive allele for the plant being short. Use the genetic diagram to calculate the probability (as a percentage) that the offspring of two Tt plants would be tall.

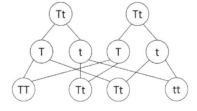

[3 marks]

Identify the tall genotypes: TT and Tt [1 mark]

Calculate the probability: Four potential genotypes. Three are TT or Tt. [1 mark]

$\frac{3}{4}$ probability the plant will be tall.

Turn probability into percentage: $\frac{3}{4} \times 100 = 75\%$ [1 mark]

Q2. The dominant allele for a plant being tall is represented by T, while t represents the recessive allele for the plant being short. Use each genetic diagram to calculate the probability (as a percentage) that the plant will be short:

a)

```
   (Tt)          (tt)
  /    \        /    \
(T)    (t)    (t)    (t)
  \     \  \  /  /    /
(Tt)    (tt)  (Tt)    (tt)
```

b)

```
   (TT)          (tt)
  /    \        /    \
(T)    (T)    (t)    (t)
  \     \  \  /  /    /
(Tt)    (Tt)  (Tt)    (Tt)
```

[3 marks each]

a) ....

....

b) ....

....

Hard questions: Need to construct own genetic diagrams. **HT**

Q3. Cystic fibrosis is caused by a recessive allele. One parent is a heterozygous carrier of cystic fibrosis, the second parent is homozygous and does not have cystic fibrosis. Draw a genetic diagram to predict the parents' offspring and calculate the percentage chance of their offspring being a carrier of cystic fibrosis. [4 marks]

....

....

Q4. Polydactyly is caused by a dominant allele. Two heterozygous parents both have polydactyly. Draw a genetic diagram to predict the parents' offspring and calculate the percentage chance of their offspring not having polydactyly. [4 marks]

....

....

[Total marks   / 22]

Using quadrats (combined science)

Quadrats are used to estimate the number of organisms in a large area, without needing to count them all.

You need to remember this!

A quadrat is generally made of a square frame of a known area (e.g. 1 m^2). The quadrat is placed at a random position within an area and the number of a certain organism counted. To estimate the total number of organisms you multiply the average number found per square metre by the overall area.

Basic questions: Areas and means given.

Model example: A garden has an area of 200 m^2. Students use a quadrat of area 1 m^2 to estimate the number of daisies in the garden. There is an average of 15 daisies per quadrat when it is placed at random positions in the garden. Estimate the total number of daisies in the garden. [1 mark]

Multiply average per m^2 by the total area: 15 × 200 = 3000 daisies [1 mark]

Q1. A field has an area of 3000 m^2. Students use a quadrat of area 1 m^2 to estimate the number of buttercups in the field. Estimate the total number of buttercups for each average number per quadrat:

a) 8   b) 9   c) 4   d) 14   [1 mark each]

a) ....

b) ....

c) ....

d) ....

Q2. Some quadrats are placed at random positions in a field containing some dandelions. There are an average of 20 dandelions per quadrat. Estimate the total number of dandelions in the field for each area:

a) 4000 m^2 b) 2500 m^2 c) 7500 m^2 d) 1800 m^2 [1 mark each]

a) ....

b) ....

c) ....

d) ....

Medium questions: Calculation of mean needed.

Q3. To estimate the number of daisies in a field, some students placed a 1 m^2 quadrat at five random positions in the field.

The results are shown in the table.

The field has an area of 4000 m^2.

Quadrat number	Number of daisies
1	12
2	15
3	18
4	10
5	20

a) Calculate the mean number of daisies per quadrat. [1 mark]

....

b) Estimate the total number of daisies in the field. [1 mark]

....

c) A nearby field gets more sunlight and there are twice as many daisies in each quadrat. If that field has an area of 10 000 m^2, estimate the number of daisies in it. Give your answer in standard form. [3 marks]

....

....

....

Hard questions: Calculation of mean and area needed.

Q4. A diagram of an 'L'-shaped garden is shown.

Five 1 m^2 quadrats are placed at random positions in the garden and the number of cornflowers per quadrat is counted.

a) Use the results table to estimate the number of cornflowers in the garden. [3 marks]

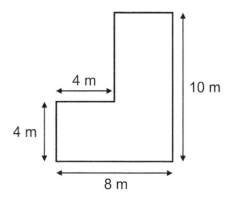

Quadrat number	Number of cornflowers
1	8
2	4
3	12
4	20
5	0

....

....

....

....

b) A shed of 4 m^2 is placed into the garden. Estimate how many cornflowers now remain in the garden. [2 marks]

....

....

[Total marks   / 18]

Efficiency of biomass transfer (biology only)

Different levels of a food chain are called trophic levels. Biomass is lost between trophic levels. In a healthy ecosystem, around 10% of the biomass is transferred from a lower trophic level to a higher one.

You need to remember this equation!

The equation to calculate efficiency of biomass transfer (as a percentage) is:

$$efficiency\ of\ biomass\ transfer = \frac{biomass\ of\ higher\ trophic\ level}{biomass\ of\ lower\ trophic\ level} \times 100$$

Basic questions: Percentage calculations.

Model example: A rabbit gains 18 g of mass after eating 200 g of grass. Calculate the efficiency of biomass transfer between the grass and the rabbit. [2 marks]

Write down the equation:

$$efficiency\ of\ biomass\ transfer = \frac{biomass\ of\ higher\ trophic\ level}{biomass\ of\ lower\ trophic\ level} \times 100$$

Substitute the variables into the equation: $efficiency\ of\ biomass\ transfer = \frac{18}{200} \times 100$ [1 mark]

Calculate answer: $efficiency\ of\ biomass\ transfer = 9\%$ [1 mark]

Q1. A fox gains 200 g of mass from eating a rabbit. Calculate, as a percentage, the efficiency of biomass transfer for each mass of rabbit eaten:

a) 2200 g **b)** 1800 g **c)** 2000 g **d)** 2400 g [2 marks each]

a) ....

b) ....

c) ....

d) ....

Q2. A lion gains 4 kg of mass from eating an antelope. Calculate, as a percentage, the efficiency of biomass transfer for each mass of antelope eaten:

a) 40 kg **b)** 43 kg **c)** 30 kg **d)** 50 kg [2 marks each]

a) ....

b) ....

c) ....

d) ....

Medium questions: Rearranging or unit conversion needed (hint in box).

Q3. A cat eats part of a bird. The efficiency of biomass transfer from the cat to the bird is 8%. Calculate how much mass the cat would gain for each mass of bird eaten:

a) 100 g   **b)** 80 g   **c)** 120 g   **d)** 30 g   [2 marks each]

a) ....

b) ....

c) ....

d) ....

Q4. A hyena eats 5 kg of wildebeest. Calculate the efficiency of biomass transfer for each mass that the hyena gains:

$g \rightarrow kg \div 1000$

a) 400 g   **b)** 380 g   **c)** 550 g   **d)** 450 g   [3 marks each]

a) ....

b) ....

c) ....

d) ....

Hard questions: Calculations include mass used in respiration and mass of excretion.

Q5. An elephant eats 160 kg of vegetation. The elephant excretes 120 kg of mass in gas, faeces and urine and uses 20 kg of mass in respiration. The remaining mass leads to an increase in mass of the elephant. Calculate the efficiency of biomass transfer.   [3 marks]

....

....

Q6. A mongoose eats 0.2 kg of a snake. The mongoose uses 40 g of mass in respiration and excretes 145 g of mass in gas faeces and urine. The remaining mass leads to an increase in mass of the mongoose. Calculate the efficiency of biomass transfer.   [4 marks]

....

....

Q7. An owl eats 0.15 kg of a mouse. The owl excretes 90 g of mass in gas, faeces and urine. The efficiency of biomass transfer is 8%. The remaining mass is used in respiration. Calculate how much mass is used in respiration.   [4 marks]

....

....

[Total marks   / 47]

Chemistry summary sheet

In the chemistry examinations, 20% of the overall marks are awarded for mathematical skills. This includes using the equations on the following page, and the general skills that you saw previously. As stated previously, examples of this include calculation of means, percentages and gradients; as well as rearrangement of equations.

The ten topics in chemistry are:

1. Atomic structure and the periodic table

2. Bonding, structure, and the properties of matter

3. Quantitative chemistry

4. Chemical changes

5. Energy changes

6. The rate and extent of chemical change

7. Organic chemistry

8. Chemical analysis

9. Chemistry of the atmosphere

10. Using resources

The equations on the following page are split across these topics and need to be remembered. Cover up the right-hand side and regularly test yourself to check that you can remember the equations.

A copy of the periodic table is supplied in chemistry examinations, and you will need to be able to use the table to answer many of the questions in this section.

The equations in **bold** are chemistry only.

Relative atomic mass	$$\text{relative atomic mass} = \frac{\text{sum of (isotope mass number} \times \text{isotope abundance)}}{100}$$
Surface area to volume ratio	$$\text{surface area} \div \text{volume}$$
Percentage mass	$$\text{percentage mass of an element in a compound} =$$ $$\frac{A_r \times \text{number of atoms of that element}}{M_r \text{ of compound}} \times 100$$
Moles	$$\text{number of moles} = \frac{\text{mass of substance (g)}}{\text{relative formula mass (g/mol)}}$$
Concentration	$$\text{concentration (g/dm}^3) = \frac{\text{mass of solute (g)}}{\text{volume of solution (dm}^3)}$$ $$\text{concentration (mol/dm}^3) = \frac{\text{number of moles of solute (mol)}}{\text{volume of solution (dm}^3)}$$
Volume of gases	$$\textbf{volume of gas (dm}^3\textbf{)} = 24 \times \frac{\text{mass of gas (g)}}{M_r \text{ of gas}}$$
Percentage yield	$$\textbf{\% yield} = \frac{\text{mass of product actually made}}{\text{maximum theoretical mass of product}} \times \textbf{100}$$
Atom economy	$$\textbf{percentage atom economy} =$$ $$\frac{\text{relative formula mass of desired product from equation}}{\text{sum of relative formula masses of all reactants from equation}} \times \textbf{100}$$
Bond energy	$$\text{energy change} =$$ $$(\text{sum of bond energies of reactants}) - (\text{sum of bond energies of products})$$
Rate of reaction	$$\text{mean rate of reaction} = \frac{\text{quantity of reactant used}}{\text{time taken}}$$ $$\text{mean rate of reaction} = \frac{\text{quantity of product formed}}{\text{time taken}}$$
Chromatography	$$R_f = \frac{\text{distance moved by substance}}{\text{distance moved by solvent}}$$

Relative formula and atomic mass (combined science)

The relative formula mass can be calculated by adding together the individual relative atomic masses of each atom in the molecule.

You need to remember this equation!

Each element has different isotopes which have different abundances. Isotopes of the same element have the same atomic number, but different mass numbers. In other words, isotopes have the same number of protons but different numbers of neutrons. The equation to calculate relative atomic mass is:

$$relative\ atomic\ mass = \frac{sum\ of\ (isotope\ mass\ number \times isotope\ abundance)}{100}$$

Give answers on this worksheet to three significant figures.

Basic questions: Relative formula mass.

Model example: Calculate the relative formula mass of H_2O [2 marks]

Substitute atomic masses into formula: $(2 \times 1) + 16$ [1 mark]

Calculate answer: $(2 \times 1) + 16 = 18$ [1 mark]

Q1. Calculate the relative formula mass for each formula:

a) O_2 b) N_2 c) HCl d) KNO_3

e) Fe_2O_3 f) H_2SO_4 g) $Ca(OH)_2$ h) $Pb(NO_3)_2$ [2 marks each]

a) ....

b) ....

c) ....

d) ....

e) ....

f) ....

g) ....

h) ....

Medium questions: Calculating relative atomic mass of two isotopes.

Model example: Lithium is found as two stable isotopes. The isotope ^6Li has an abundance of 7.6% and ^7Li has an abundance of 92.4%. Calculate the relative atomic mass of lithium. [2 marks]

Write down the equation: $relative\ atomic\ mass = \frac{sum\ of\ (isotope\ mass\ number \times isotope\ abundance)}{100}$

Substitute the variables into the equation: $= \frac{(6 \times 7.6) + (7 \times 92.4)}{100}$ [1 mark]

Calculate answer: $\frac{(6 \times 7.6) + (7 \times 92.4)}{100} = 6.92$ [1 mark]

Q2. Chlorine is found as two stable isotopes. The isotope ^{35}Cl has an abundance of 76% and ^{37}Cl has an abundance of 24%. Calculate the relative atomic mass of chlorine. [2 marks]

....

....

Q3. Boron is found as two stable isotopes. The isotope ^{10}B has an abundance of 20% and ^{11}B has an abundance of 80%. Calculate the relative atomic mass of boron. [2 marks]

....

....

Q4. Copper is found as two stable isotopes. The isotope ^{63}Cu has an abundance of 69.2% and ^{65}Cu has an abundance of 30.8%. Calculate the relative atomic mass of copper. [2 marks]

....

....

Q5. Gallium is found as two stable isotopes. The isotope ^{69}Ga has an abundance of 60.1% and ^{70}Ga has an abundance of 39.9%. Calculate the relative atomic mass of gallium. [2 marks]

....

....

Q6. Bromine is found as two stable isotopes. The isotope ^{79}Br has an abundance of 51% and ^{81}Br has an abundance of 49%. Calculate the relative atomic mass of bromine. [2 marks]

....

....

Hard questions: Calculating relative atomic mass of three isotopes.

Q7. Neon is found as three stable isotopes. The isotope ^{20}Ne has an abundance of 90.48%, ^{21}Ne has an abundance of 0.27% and ^{22}Ne has an abundance of 9.25%. Calculate the relative atomic mass of neon. [2 marks]

....

....

Q8. Magnesium is found as three stable isotopes. The isotope ^{24}Mg has an abundance of 79%, ^{25}Mg has an abundance of 10% and ^{26}Mg has an abundance of 11%. Calculate the relative atomic mass of magnesium. [2 marks]

....

....

[Total marks   / 30]

Bulk and surface properties of matter (chemistry only)

There are three categories of particles. Which category a particle is in depends on its diameter:

> **You need to remember this!**

- *Nanoparticles* have a diameter between 1 and 100 nanometres. Given that a single atom has a diameter of ~0.1 nm, most nanoparticles only consist of several hundred atoms.
- *Fine particles* (PM$_{2.5}$) have a diameter between 100 and 2500 nanometres.
- *Coarse particles* (PM$_{10}$) have a diameter between 2500 and 10 000 nanometres and are also commonly called dust.

The smaller the particle size, the greater the surface area to volume ratio. Nanoparticles have particularly large surface area to volume ratios.

The equation for surface area to volume ratio is: *surface area ÷ volume*

Basic questions: No rearranging or unit conversion needed.

Model example: A nanoparticle has a surface area of 15 000 nm^2 and a volume of 125 000 nm^3. Calculate the surface area to volume ratio for the nanoparticle. **[3 marks]**

Write down equation: *surface area ÷ volume*

Substitute variables into equation: *surface area ÷ volume* = 15 000 ÷ 125 000   **[1 mark]**

Calculate answer: 15 000 ÷ 125 000 = 0.12   **[1 mark]**

Write units: 0.12 nm^{-1} **[1 mark]**

Q1. A nanoparticle has a surface area of 8000 nm^2. Calculate the surface area to volume ratio for each volume:

     a) 80 000 nm^3 **b)** 100 000 nm^3 **c)** 140 000 nm^3 **d)** 60 000 nm^3 **[3 marks each]**

     a) ....

     b) ....

     c) ....

     d) ....

Medium questions: Rearranging needed.

Q2. A nanoparticle has a volume of 25 000 nm^3. Calculate the surface area of the nanoparticle for each surface area to volume ratio:

     a) 0.1 nm^{-1} **b)** 0.08 nm^{-1} **c)** 0.40 nm^{-1} **d)** 0.45 nm^{-1} **[3 marks each]**

a) ....

....

b) ....

....

c) ....

....

d) ....

....

Hard questions: Need to calculate surface area and volume individually.

Model example: Calculate the surface area to volume ratio of a nanoparticle that is in the shape of a cube of side length 20 nm. [4 marks]

Calculate the surface area of the cube:	each side = 20 × 20 = 400 nm²

six sides overall, so total surface area
= 6 × 400 = 2400 nm² [1 mark]

Calculate the volume of the cube: *volume* = 20 × 20 × 20 = 8000 nm³ [1 mark]

Calculate the surface area to volume ratio of the cube: *surface area* ÷ *volume* = 2400 ÷ 8000 = 0.3 [1 mark]

Write units: 0.3 nm⁻¹ [1 mark]

Q3. Calculate the surface area to volume ratio of a cubic nanoparticle of side length:

a) 1 nm   **b)** 5 nm   **c)** 15 nm   **d)** 4 nm   [4 marks each]

a) ....

....

b) ....

....

c) ....

....

d) ....

....

[Total marks   / 40]

Percentage mass (combined science)

To calculate the percentage of mass of an element in a compound, the following equation can be used:

Percentage mass of an element in a compound $= \frac{A_r \times number\ of\ atoms\ of\ that\ element}{M_r\ of\ compound} \times 100$

Here, A_r is the relative atomic mass of the element and M_r is the formula mass.

You need to remember this equation!

Basic questions: Molecular formulae with two elements.

Model example: Calculate the percentage by mass of copper in copper oxide, CuO [2 marks]

Substitute variables into equation: *percentage mass of an element in a compound* [1 mark]

$$= \frac{1 \times 63.5}{79.5} \times 100$$

Calculate answer: $\frac{1 \times 63.5}{79.5} \times 100 = 80\%$ [1 mark]

Q1. Calculate the percentage by mass of carbon in carbon monoxide, CO [2 marks]

....

Q2. Calculate the percentage by mass of hydrogen in hydrochloric acid, HCl [2 marks]

....

Q3. Calculate the percentage by mass of chlorine in potassium chloride, KCl [2 marks]

....

Q4. Calculate the percentage by mass of oxygen in magnesium oxide, MgO [2 marks]

....

Q5. Calculate the percentage by mass of sodium in sodium chloride, NaCl [2 marks]

....

Q6. Calculate the percentage by mass of bromine in copper bromide, CuBr [2 marks]

....

Medium questions: Complex molecular formulae.

Model example: Calculate the percentage by mass of hydrogen in sulfuric acid, H_2SO_4 [3 marks]

Calculate formula mass: $(1 \times 2) + 32 + (16 \times 4) = 98$ [1 mark]

Substitute variables into equation: *percentage mass of an element in a compound* [1 mark]

$$= \frac{1 \times 2}{98} \times 100$$

Calculate answer: $\frac{1 \times 2}{98} \times 100 = 2.0\%$ [1 mark]

Q7. Calculate the percentage by mass of carbon in carbon dioxide, CO_2 [3 marks]

....

....

Q8. Calculate the percentage by mass of sulfur in sulfur dioxide, SO_2 [3 marks]

....

....

Q9. Calculate the percentage by mass of hydrogen in benzene, C_6H_6 [3 marks]

....

....

Q10. Calculate the percentage by mass of copper in copper sulfate, $CuSO_4$ [3 marks]

....

....

Q11. Calculate the percentage by mass of magnesium in magnesium carbonate, $MgCO_3$ [3 marks]

....

....

Hard questions: More complex molecular formulae.

Q12. Calculate the percentage by mass of aluminium in aluminium sulfate, $Al_2(SO_4)_3$ [3 marks]

....

....

Q13. Calculate the percentage by mass of nitrogen in magnesium nitrate, $Mg(NO_3)_2$ [3 marks]

....

....

Q14. Calculate the percentage by mass of oxygen in copper carbonate, $Cu_2CO_3(OH)_2$ [3 marks]

....

....

Q15. Calculate the percentage by mass of phosphorus in ammonium phosphate, $(NH_4)_3PO_4$ [3 marks]

....

....

[Total marks   / 39]

Moles (combined science) HT

If you have one mole of a substance, it means that you have 6.02×10^{23} particles of that substance. That number is known as the Avogadro constant. The mass (in grams) of one mole of a substance is equal to its relative formula mass, M_r. To calculate the number of moles of an element, you use relative atomic mass, A_r instead of the relative formula mass, M_r.

To calculate the number of moles of a substance, you can use the equation:

$$number\ of\ moles = \frac{mass\ of\ substance\ (g)}{relative\ formula\ mass\ (g/mol)}$$

You need to remember this equation!

Give answers on this worksheet to three significant figures.

Basic questions: Calculations involving elements.

Model example: Calculate the number of moles in 96 g of titanium.   [1 mark]

Write down the equation: $number\ of\ moles = \frac{mass\ of\ substance\ (g)}{relative\ formula\ mass\ (g/mol)}$

Substitute the variables into the equation: $\frac{96}{48} = 2$ mol   [1 mark]

Q1. Calculate the number of moles in:

  a) 20 g of calcium   **b)** 460 g of sodium   **c)** 96 g of sulfur   **d)** 0.35 g of lithium   [1 mark each]

  a) ....

  b) ....

  c) ....

  d) ....

Q2. Calculate the mass of the following elements:

  a) 3 moles of iron   **b)** 2 moles of carbon **c)** 0.2 moles of tin   **d)** 0.05 moles of zinc [1 mark each]

  a) ....

  b) ....

  c) ....

  d) ....

Q3. Calculate the relative atomic mass of the following elements. Use this to identify the element:

  a) 2 moles of this element have a mass of 150 g

  b) 10 moles of this element have a mass of 560 g

  c) 5 moles of this element have a mass of 480 g

  d) 0.5 moles of this element have a mass of 19.5 g   [2 marks each]

  a) ....

  b) ....

  c) ....

  d) ....

Medium questions: Calculations involving compounds.

Model example: Calculate the number of moles in 0.32 g of Fe_2O_3 [2 marks]

Write down the equation:

$$number\ of\ moles = \frac{mass\ of\ substance\ (g)}{relative\ formula\ mass\ (g/mol)}$$

Calculate the relative formula mass: $(2 \times 56) + (3 \times 16) = 160$ [1 mark]

Substitute the variables into the equation: $\frac{0.32}{160} = 0.002$ mol [1 mark]

Q4. Calculate the number of moles in:

a) 80 g of O_2 **b)** 4 g of NaOH   **c)** 204 g of Al_2O_3 **d)** 74 g of $Mg(NO_3)_2$ [2 marks each]

a) ....

b) ....

c) ....

d) ....

Q5. Calculate the mass of:

a) 0.6 moles of H_2O **c)** 4 moles of NH_4Cl

b) 2 moles of $Mg(OH)_2$ **d)** 0.03 moles of $CaCO_3$ [2 marks each]

a) ....

b) ....

c) ....

d) ....

Hard questions: Calculations involving compounds. Requires unit conversion (hints in boxes).

Q6. Calculate the number of moles in 4.59 kg of Al_2O_3 | kg → g × 1000 | [3 marks]

....

....

Q7. Calculate the number of moles in 2.12 mg of Na_2CO_3 | mg → g ÷ 1000 |
Give your answer in standard form.   [3 marks]

....

....

Q8. Calculate the number of moles in 1.89 kg of HNO_3 [3 marks]

....

....

[Total marks   / 41]

Balancing symbol equations (combined science)

In a chemical reaction, the total mass of the reactants must be the same as the total mass of the products.

You need to remember this!

A symbol equation must always be balanced. This means that there must be the same number of atoms of each type of element on both sides of the equation.

Basic questions: Only one term in symbol equation needs balancing.

Model example: Balance the symbol equation $H_2 + O_2 \rightarrow 2H_2O$ [1 mark]

Count the number of each element on both sides of the equation:

On left: 2 H and 2 O atoms.

On right: 4 H and 2 O atoms.

Add any numbers required before the front of the chemical formulae to balance:

$2H_2 + O_2 \rightarrow 2H_2O$ [1 mark]

Q1. Balance each of the symbol equations:

a) $2Na + Cl_2 \rightarrow NaCl$

b) $Zn + HCl \rightarrow ZnCl_2 + H_2$

c) $2Mg + O_2 \rightarrow MgO$

d) $Al + 3O_2 \rightarrow 2Al_2O_3$

e) $2K + 2H_2O \rightarrow KOH + H_2$

f) $NaOH + H_2SO_4 \rightarrow 2H_2O + Na_2SO_4$ [1 mark each]

a) ....

b) ....

c) ....

d) ....

e) ....

f) ....

Medium questions: Balancing multiple terms.

Q2. Balance each of the symbol equations:

a) $H_2O_2 \rightarrow O_2 + H_2O$

b) $N_2 + H_2 \rightarrow NH_3$

c) $Na + H_2SO_4 \rightarrow Na_2SO_4 + H_2$

d) $CO_2 + H_2O \rightarrow C_6H_{12}O_6 + O_2$

e) $CuCO_3 \rightarrow CuO + CO_2$

f) $C_6H_6 + O_2 \rightarrow CO_2 + H_2O$ [2 marks each]

a) ....

b) ....

c) ....

d) ....

e) ....

f) ....

Hard questions: Balancing equations using reacting masses. **HT**

Model example: A mass of 200 g of calcium carbonate ($CaCO_3$) is heated and undergoes thermal decomposition to form 122 g of calcium oxide (CaO) and 88 g of carbon dioxide (CO_2). Write a balanced symbol equation for the thermal decomposition of calcium carbonate. [3 marks]

Calculate the relative formula mass of each substance:	$CaCO_3$: 40 + 12 + (3 × 16) = 100
	CO_2: 12 + (2 × 16) = 44
	CaO: 40 + 16 = 56 [1 mark]
Calculate the number of moles of each substance:	$CaCO_3$: 200 ÷ 100 = 2 moles
	CO_2: 88 ÷ 44 = 2 moles
	CaO: 122 ÷ 56 = 2 moles [1 mark]
Simplify ratio and write balanced symbol equation:	Ratio 1:1:1
	$CaCO_3 \rightarrow CaO + CO_2$ [1 mark]

Q3. In a displacement reaction, 130 g of zinc (Zn) reacts completely with 319 g of copper sulfate ($CuSO_4$) to form 127 g of copper and 322 g of zinc sulfate ($ZnSO_4$). Write a balanced symbol equation for this displacement reaction. [3 marks]

....

....

....

....

Q4. 39 g of potassium (K) reacts completely with 18 g of water (H_2O) to form 56 g of potassium hydroxide (KOH) and 1 g of hydrogen (H_2). Write a balanced symbol equation for this reaction. [3 marks]

....

....

....

....

[Total marks   / 24]

Concentration of solutions (combined science)

Concentration is a measure how much of a substance is in a certain volume. In combined science, concentration (in units of g/dm³) can be calculated using:

$$concentration \; (\text{g/dm}^3) = \frac{mass \; of \; solute \; (\text{g})}{volume \; of \; solution \; (\text{dm}^3)}$$

HT It can also be calculated (chemistry only) in units of mol/dm³ by using:

$$concentration \; (\text{mol/dm}^3) = \frac{number \; of \; moles \; of \; solute \; (\text{mol})}{volume \; of \; solution \; (\text{dm}^3)}$$

You need to remember these equations!

Basic questions: Concentration of solutions.

Model example: Calculate the concentration of 80 g of sodium chloride that dissolves to make 40 dm³ of solution. **[3 marks]**

Write down equation:
$$concentration \; (\text{g/dm}^3) = \frac{mass \; of \; solute \; (\text{g})}{volume \; of \; solution \; (\text{dm}^3)}$$

Substitute variables into equation: $concentration = 80 \div 40$ [1 mark]

Calculate answer: $concentration = 80 \div 40 = 2$ [1 mark]

Write units: 2 g/dm³ [1 mark]

Q1. Calculate the concentration in grams per dm³ of 20 g of sodium chloride dissolved in water to make each volume of solution:

a) 8 dm³ b) 40 dm³ c) 0.5 dm³ d) 3 dm³ [3 marks each]

a) ....

b) ....

c) ....

d) ....

Q2. Calculate the concentration of 50 g of copper sulfate dissolved in water to make each volume of solution:

a) 250 dm³ b) 800 dm³ c) 1200 dm³ d) 20 dm³ [3 marks each]

a) ....

b) ....

c) ....

d) ....

Medium questions: Concentration of solutions. Rearranging needed.

Q3. A sports drink has a glucose concentration of 0.1 g/dm^3. Calculate the mass of glucose in each volume of sports drink:

a) 0.5 dm^3 **b)** 0.2 dm^3 **c)** 0.75 dm^3 **d)** 0.04 dm^3 [3 marks each]

a) ....

b) ....

c) ....

d) ....

Q4. A limewater solution has a calcium hydroxide concentration of 0.8 g/dm^3. Calculate the volume of solution for each calcium hydroxide mass:

a) 1 g   **b)** 0.5 g   **c)** 0.2 g   **d)** 1.5 g   [3 marks each]

a) ....

b) ....

c) ....

d) ....

Hard questions: Titration calculations (chemistry only). **HT**

Q5. A student titrates 0.05 dm^3 of 0.6 mol/dm^3 sodium hydroxide solution to neutralise 0.075 dm^3 of sulfuric acid. The equation for the reaction is $2NaOH + H_2SO_4 \rightarrow Na_2SO_4 + 2H_2O$

a) Calculate the number of moles of sodium hydroxide titrated.   [1 mark]

....

b) State the number of moles of sulfuric acid neutralised.   [1 mark]

....

c) Calculate the concentration of the sulfuric acid.   [3 marks]

....

Q6. A student titrates 0.04 dm^3 of 0.01 mol/dm^3 potassium hydroxide solution to neutralise some hydrochloric acid of 0.04 mol/dm^3 concentration. The equation for the reaction is $HCl + KOH \rightarrow KCl + H_2O$

a) Calculate the number of moles of potassium hydroxide titrated.   [1 mark]

....

b) State the number of moles of hydrochloric acid neutralised.   [1 mark]

....

c) Calculate the volume of the hydrochloric acid.   [3 marks]

....

[Total marks   / 58]

Volumes of gases (chemistry only) HT

At room temperature (20 °C) and pressure (1 atm), one mole of gas occupies a volume of 24 dm^3. The volume of a gas at room temperature and atmospheric pressure can be calculated by using:

$$volume\ of\ gas\ (\text{dm}^3) = 24 \times \frac{mass\ of\ gas\ (\text{g})}{M_r\ of\ gas}$$

You need to remember this equation!

Basic questions: Relative formula mass given.

Model example: 64 g of oxygen (O_2) is at room temperature and atmospheric pressure. The relative formula mass of O_2 is 32. Calculate the volume of oxygen. [3 marks]

Write down equation: $volume\ of\ gas\ (\text{dm}^3) = 24 \times \frac{mass\ of\ gas\ (\text{g})}{M_r\ of\ gas}$

Substitute variables into equation: $volume\ of\ gas\ (\text{dm}^3) = 24 \times \frac{64}{32}$ [1 mark]

Calculate answer: $volume\ of\ gas\ (\text{dm}^3) = 24 \times \frac{64}{32} = 48$ [1 mark]

Write units: 48 dm^3 [1 mark]

Q1. Some nitrogen (N_2) gas is at room temperature and atmospheric pressure. The relative formula mass of N_2 is 28. Calculate the volume of nitrogen for each mass:

a) 14 g b) 21 g c) 40 g d) 5 g [3 marks each]

a) ....

b) ....

c) ....

d) ....

Medium questions: Calculation of relative formula mass needed.

Model example: Calculate the volume of 88 g of carbon dioxide (CO_2) gas at room temperature and atmospheric pressure. [4 marks]

Write down equation: $volume\ of\ gas\ (\text{dm}^3) = 24 \times \frac{mass\ of\ gas\ (\text{g})}{M_r\ of\ gas}$

Calculate M_r of the gas. $M_r = 12 + (2 \times 16) = 44$ [1 mark]

Substitute variables into equation: $volume\ of\ gas\ (\text{dm}^3) = 24 \times \frac{88}{44}$ [1 mark]

Calculate answer: $volume\ of\ gas\ (\text{dm}^3) = 24 \times \frac{88}{44} = 48$ [1 mark]

Write units: 48 dm^3 [1 mark]

Q2. Calculate the volume of 40 g of hydrogen (H_2) gas at room temperature and atmospheric pressure. [4 marks]

....

....

Q3. Calculate the volume of 50 g of sulfur dioxide (SO_2) gas at room temperature and atmospheric pressure. [4 marks]

....

....

Q4. Calculate the volume of 64 g of methane (CH_4) gas at room temperature and atmospheric pressure. [4 marks]

....

....

Q5. Calculate the volume of 220 g of butane (C_4H_{10}) gas at room temperature and atmospheric pressure. [4 marks]

....

....

Hard questions: Rearranging and unit conversion needed.

Q6. Calculate the mass of chlorine (Cl_2) gas that has a volume of 200 cm^3 at room temperature and atmospheric pressure. [5 marks]

....

....

Q7. Calculate the mass of carbon monoxide (CO) gas that has a volume of 720 cm^3 at room temperature and atmospheric pressure. [5 marks]

....

....

Q8. Calculate the mass of propane (C_3H_8) gas that has a volume of 6000 cm^3 at room temperature and atmospheric pressure. [5 marks]

....

....

[Total marks   / 43]

Percentage yield (chemistry only)

Every reaction has a maximum theoretical product produced from given reactants. The percentage yield is a measure of how much of this maximum theoretical yield was actually produced. The equation to calculate percentage yield is:

You need to remember this equation!

$$\% \, yield = \frac{mass \ of \ product \ actually \ made}{maximum \ theoretical \ mass \ of \ product} \times 100$$

Basic questions: No rearranging or unit conversion needed.

Model example: The theoretical mass of product in a reaction is 1500 g. Calculate the percentage yield if 1200 g of product is actually made. [2 marks]

Write down equation:

$$\% \, yield = \frac{mass \ of \ product \ actually \ made}{maximum \ theoretical \ mass \ of \ product} \times 100$$

Substitute variables into equation:

$$\% \, yield = \frac{1200}{1500} \times 100$$ [1 mark]

Calculate answer:

$$\% \, yield = \frac{1200}{1500} \times 100 = 80\%$$ [1 mark]

Q1. The theoretical mass of product in a reaction is 2000 g. Calculate the percentage yield for each mass of product actually made:

a) 1200 g   b) 1500 g   c) 100 g   d) 500 g   [2 marks each]

a) ....

b) ....

c) ....

d) ....

Q2. The theoretical mass of product in a reaction is 400 g. Calculate the percentage yield for each mass of product actually made:

a) 60 g   b) 220 g   c) 300 g   d) 20 g   [2 marks each]

a) ....

b) ....

c) ....

d) ....

Medium questions: Rearranging needed.

Q3. The percentage yield of a reaction is 60%. Calculate the mass of the product actually made for each maximum theoretical product mass:

a) 2000 g   b) 500 g   c) 800 g   d) 40 g   [3 marks each]

a) ....

b) ....

c) ....

d) ....

Q4. The percentage yield of a reaction is 20%. Calculate the maximum theoretical product mass for each product mass actually made:

a) 400 g   **b)** 25 g   **c)** 30 g   **d)** 220 g   [3 marks each]

a) ....

b) ....

c) ....

d) ....

Hard questions: Need to calculate maximum theoretical mass of product. HT

Q5. A student used 6.5 g of zinc in the following reaction: $Zn + CuSO_4 \rightarrow ZnSO_4 + Cu$

    a) Use the symbol equation to calculate the maximum mass of zinc sulfate ($ZnSO_4$) that could be produced.   [4 marks]

    ....

    ....

    b) The reaction produced 6 g of zinc sulfate. Using your answer to part a), calculate the percentage yield of this reaction.   [2 marks]

    ....

    ....

Q6. A student used 127 g of copper in the following reaction: $Cu + 2H_2SO_4 \rightarrow CuSO_4 + SO_2 + 2H_2O$

    a) Use the symbol equation to calculate the maximum mass of copper sulfate ($CuSO_4$) produced.   [4 marks]

    ....

    ....

    b) The reaction produced 256 g of copper sulfate. Using your answer to part a), calculate the percentage yield of this reaction.   [2 marks]

    ....

    ....

[Total marks   / 52]

Atom economy (chemistry only)

The atom economy is a measure of how much mass of the reactants gets turned into the desired products. The equation to calculate the percentage atom economy is:

$$percentage\ atom\ economy = \frac{relative\ formula\ mass\ of\ desired\ product\ from\ equation}{sum\ of\ relative\ formula\ masses\ of\ all\ reactants\ from\ equation} \times 100$$

You need to remember this equation!

Basic questions: No rearranging or unit conversion needed.

Model example: The relative formula mass of desired product is 32. The sum of relative formula masses of all reactants is 64. Calculate the percentage atom economy. [2 marks]

Substitute variables into equation: $percentage\ atom\ economy = \frac{32}{64} \times 100$ [1 mark]

Calculate answer: $percentage\ atom\ economy = \frac{32}{64} \times 100 = 50\%$ [1 mark]

Q1. The relative formula mass of desired product is 12. Calculate the percentage atom economy for each sum of relative formula masses of all reactants:

a) 24   **b)** 30   **c)** 36   **d)** 40   [2 marks each]

a) ....

b) ....

c) ....

d) ....

Q2. The relative formula mass of desired product is 20. Calculate the percentage atom economy for each sum of relative formula masses of all reactants:

a) 25   **b)** 50   **c)** 80   **d)** 22   [2 marks each]

a) ....

b) ....

c) ....

d) ....

Medium questions: Rearranging needed.

Q3. The percentage atom economy of a reaction is 80%. Calculate the sum of the relative formula masses of the reactants for each relative formula mass of the desired product.

a) 80 g   **b)** 120 g   **c)** 15 g   **d)** 25 g   [3 marks each]

a) ....

b) ....

c) ....

d) ....

Hard questions: Calculation of relative formula masses needed.

Model example: Calculate the percentage atom economy of extracting Iron (Fe) from iron oxide:

$Fe_2O_3 + 3CO \rightarrow 2Fe + 3CO_2$ [4 marks]

Calculate M_r of desired product:	$2 \times 56 = 112$	[1 mark]
Calculate M_r of all reactants:	$(2 \times 56) + (3 \times 16) + 3 \times (12 + 16) = 244$	[1 mark]
Substitute variables into equation:	$percentage\ atom\ economy = \frac{112}{244} \times 100$	[1 mark]
Calculate answer:	$percentage\ atom\ economy = \frac{112}{244} \times 100 = 46\%$	[1 mark]

Q4. Electrolysis can be used to extract hydrogen from water. Calculate the percentage atom economy of extracting hydrogen (H_2) from water (H_2O): $2H_2O \rightarrow 2H_2 + O_2$ [4 marks]

....

....

....

Q5. Calculate the percentage atom economy of the production of calcium oxide (CaO) from the thermal decomposition of calcium carbonate ($CaCO_3$): $CaCO_3 \rightarrow CaO + CO_2$ [4 marks]

....

....

....

Q6. Calculate the percentage atom economy of the production of ammonia (NH_3) from ammonium sulfate $(NH_4)_2SO_4$: $(NH_4)_2SO_4 \rightarrow H_2SO_4 + 2NH_3$ [4 marks]

....

....

....

Q7. Calculate the percentage atom economy of fermentation where ethanol (C_2H_5OH) is produced from glucose ($C_6H_{12}O_6$): $C_6H_{12}O_6 \rightarrow 2C_2H_5OH + 2CO_2$ [4 marks]

....

....

....

[Total marks   / 44]

Bond energy calculations (combined science) HT

When a new bond is formed, energy is released. However, it takes a supply of energy to break a bond. The amount of energy needed to break one mole of a bond is called the bond energy. Bond energy is usually given in units of kJ/mol.

You need to remember this equation!

In a chemical reaction, the overall change in energy can be calculated by using:

energy change = (sum of bond energies of reactants) – (sum of bond energies of products)

If the energy change is negative, then that means that more energy is released than supplied. This is called an exothermic reaction. If the energy change is positive, then that means that more energy is supplied than released. This is called an endothermic reaction.

Give answers on this worksheet to four significant figures.

Basic questions: One type of bond only.

Model example: Calculate the energy needed to break all of the bonds in one mole of water (H_2O). $O - H$ bond energy = 464 kJ/mol.

[1 mark]

Count number of $O - H$ bonds:	2	
Calculate total energy needed:	2 × 464 = 928 kJ	[1 mark]

Q1. Calculate the energy needed to break all of the bonds in one mole of each of the molecules.

$C = O$ bond energy = 799 kJ/mol $\qquad$ $C - H$ bond energy = 413 kJ/mol
$H - N$ bond energy = 391 kJ/mol $\qquad$ $C - Cl$ bond energy = 346 kJ/mol

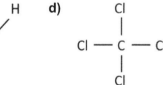

a) $O = C = O$ $\qquad$ **b)** (methane) $\qquad$ **c)** NH_3 $\qquad$ **d)** CCl_4 [1 mark each]

a) ....

b) ....

c) ....

d) ....

Medium questions: More than one type of bond.

Q2. Calculate the energy needed to break all of the bonds in one mole of each of the molecules.

$C - H$ bond energy = 413 kJ/mol $\qquad$ $C - Cl$ bond energy = 327 kJ/mol
$C - C$ bond energy = 348 kJ/mol $\qquad$ $C = C$ bond energy = 614 kJ/mol
$C - O$ bond energy = 358 kJ/mol $\qquad$ $O - H$ bond energy = 464 kJ/mol

a)

H
|
H — C — Cl
|
H

b)

H   H
| |
C = C
| |
H   H

c)

H   H
| |
H — C — C — H
| |
H   H

d)

H   H
| |
H — C — C — O — H
| |
H   H

[1 mark each]

a) ....

b) ....

c) ....

d) ....

Hard questions: Calculations of overall energy changes.

Model example: Calculate the energy change if 1 mole of hydrogen gas is combusted to form water. State whether the reaction is exothermic or endothermic.

$$2[H — H] + O = O \rightarrow 2[H — O — H]$$

[4 marks]

H — H bond energy = 436 kJ/mol   O = O bond energy = 498 kJ/mol
O — H bond energy = 464 kJ/mol

Calculate total bond energy of reactants: (2 × 436) + 498 = 1370 kJ   [1 mark]

Calculate total bond energy of products: 4 × 464 = 1856 kJ   [1 mark]

Calculate total energy change: 1370 − 1856 = 486 kJ   [1 mark]

State whether reaction is exothermic or endothermic: Energy change is negative, so this is an exothermic reaction.   [1 mark]

Q3. Calculate the energy change if 1 mole of carbon reacts with oxygen to form carbon dioxide. State whether the reaction is exothermic or endothermic.

$$C + O = O \rightarrow O = C = O$$

O = O bond energy = 498 kJ/mol   C = O bond energy = 799 kJ/mol   [4 marks]

....

....

Q4. Calculate the energy change if 1 mole of methane reacts with oxygen in the following reaction. State whether the reaction is exothermic or endothermic.

$$H — \overset{\displaystyle H}{\underset{\displaystyle H}{C}} — H + 2[O = O] \rightarrow O = C = O + 2[H — O — H]$$

C — H bond energy = 413 kJ/mol   O = O bond energy = 498 kJ/mol
C = O bond energy = 799 kJ/mol   O — H bond energy = 464 kJ/mol   [4 marks]

....

....

[Total marks   / 16]

Rates of reaction (combined science)

Different chemical reactions happen at different rates. This rate is affected by temperature, particle size, concentration and whether or not a catalyst is used.

You need to remember these equations!

You can calculate the mean rate of a reaction by using either of the following equations:

$$mean\ rate\ of\ reaction = \frac{quantity\ of\ reactant\ used}{time\ taken}$$

$$mean\ rate\ of\ reaction = \frac{quantity\ of\ product\ formed}{time\ taken}$$

The mean rate of reaction can have units of g/s, mol/s or cm^3/s.

Basic questions: No rearranging needed.

Model example: A reaction uses 240 g of reactant in a time of 60 seconds. Calculate the rate of reaction. **[3 marks]**

Write down equation:

$$mean\ rate\ of\ reaction = \frac{quantity\ of\ reactant\ used}{time\ taken}$$

Substitute variables into equation:

$$mean\ rate\ of\ reaction = \frac{240}{60}$$ **[1 mark]**

Calculate answer:

$$mean\ rate\ of\ reaction = \frac{240}{60} = 4$$ **[1 mark]**

Write units:

4 g/s **[1 mark]**

Q1. A reaction uses 400 g of reactant. Calculate the rate of reaction for each time taken:

a) 8 s   **b)** 250 s   **c)** 300 s   **d)** 25 s   **[3 marks each]**

a) ....

b) ....

c) ....

d) ....

Q2. A reaction produces 60 cm^3 of a product. Calculate the rate of reaction for each time taken:

a) 5 s   **b)** 25 s   **c)** 240 s   **d)** 4 s   **[3 marks each]**

a) ....

b) ....

c) ....

d) ....

Medium questions: Rearranging needed.

Q3. A reaction has a mean reaction rate of 20 mol/s. Calculate the quantity of product formed for each time taken:

a) 10 s   **b)** 25 s   **c)** 0.2 s   **d)** 3 s   **[3 marks each]**

a) ....

b) ....

c) ....

d) ....

Q4. A reaction has a mean reaction rate of 50 g/s. Calculate the time taken for each quantity of reactant used:

a) 125 g   **b)** 400 g   **c)** 30 g   **d)** 1800 g   [3 marks each]

a) ....

b) ....

c) ....

d) ....

Hard questions: Rates of reaction from tangents of curves. HT

Q5. Some magnesium was added to hydrochloric acid and the volume of hydrogen gas produced was measured. The graph shows the volume of hydrogen gas produced over time.

a) Which graph, A or B, shows the experiment in which the hydrochloric acid had the highest temperature? [1 mark]

....

b) Calculate the rate of reaction for graph B at a time of 0 seconds. [4 marks]

....

....

c) Calculate the rate of reaction for graph B at a time of 12 seconds. [4 marks]

....

....

[Total marks   / 57]

Chromatography (combined science)

Chromatography is used to separate substances in a mixture. There are two different "phases" in chromatography; the mobile phase and the stationary phase.

Different dissolved substances spend differing amounts of time in each phase. If a substance spends longer in the mobile phase, then it will travel further up the chromatography paper.

The R_f value is a measure of how far the dissolved substance (solute) travels compared to how far the solvent travels. The R_f value can be calculated by using the equation:

$R_f = \dfrac{distance\ moved\ by\ substance}{distance\ moved\ by\ solvent}$

> **You need to remember this equation!**

Basic questions: No rearranging needed.

Model example: A colour in a dye moves 2 cm up some chromatography paper. The solvent travels 8 cm up the chromatography paper. Calculate the R_f value for the distance travelled by the dye.   [2 marks]

Write down equation: $R_f = \dfrac{distance\ moved\ by\ substance}{distance\ moved\ by\ solvent}$

Substitute variables into equation: $R_f = \dfrac{2}{8}$ [1 mark]

Calculate answer: $R_f = \dfrac{2}{8} = 0.25$ [1 mark]

Q1. Chromatography is used to measure the R_f values for some food colouring. The solvent travels 12 cm up the chromatography paper. Calculate the R_f value for each distance travelled by the food colouring:

    a) 6 cm   **b)** 2 cm   **c)** 10 cm   **d)** 1.5 cm   [2 marks each]

    a) ....

    b) ....

    c) ....

    d) ....

Q2. A colour in a dye travels 3 cm up some chromatography paper. Calculate the R_f value for each distance moved by the solvent:

    a) 6 cm   **b)** 8 cm   **c)** 9 cm   **d)** 4 cm   [2 marks each]

    a) ....

    b) ....

    c) ....

    d) ....

Medium questions: Rearranging needed.

Q3. Chromatography is used to determine which coloured additives are in some food. The R_f value of one of the additives is 0.7. Calculate the distance moved by the additive for each distance moved by the solvent:

    a) 4 cm   **b)** 5 cm   **c)** 7 cm   **d)** 3 cm   [3 marks each]

a) ....

b) ....

c) ....

d) ....

Q4. A colour in a dye has an R_f value of 0.6. Calculate the distance moved by the solvent for each distance moved by the dye:

a) 2 cm   **b)** 3 cm   **c)** 9 cm   **d)** 1.5 cm   [3 marks each]

a) ....

b) ....

c) ....

d) ....

Hard questions: Need to measure distances using a ruler.

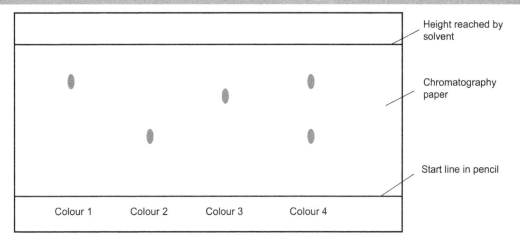

Q5. Chromatography is used to compare the R_f values of some different colours of ink.

a) Calculate the R_f value of colour 1   [3 marks]

    ....

b) Calculate the R_f value of colour 2   [3 marks]

    ....

c) Calculate the R_f value of colour 3   [3 marks]

    ....

d) State the colours that Colour 4 consists of.   [1 mark]

    ....

[Total marks   / 50]

Physics (paper 1) summary sheet

In the physics examinations, 30% of the overall marks are awarded for mathematical skills. It's therefore vital to commit the equations to memory. There are additional equations that are given on the equation sheet.

The four topics in physics paper 1 are:

1. Energy
2. Electricity
3. Particle model of matter
4. Atomic structure

The equations on the following page are split across these topics. In addition to these, remembering the units of each quantity is vital. Use the table below to quiz yourself on the units. Cover up the right-hand side and check that you can remember each symbol, unit and unit symbol.

Variables	Symbol	Units
charge	Q	coulombs (C)
current	I	amps (A)
density	ρ	kilograms per metre cubed (kg/m^3)
energy	E	joules (J)
extension	e	metres (m)
gravitational field strength	g	newtons per kilogram (N/kg)
height	h	metres (m)
mass	m	kilograms (kg)
potential difference	V	volts (V)
power	P	watts (W)
pressure	p	pascals (Pa)
resistance	R	ohms (Ω)
specific heat capacity	c	joules per kilogram per degree Celsius (J/kg °C)
specific latent heat	L	joules per kilogram (J/kg)
speed	v	metres per second (m/s)
spring constant	k	newtons per metre (N/m)
temperature	θ	degrees Celsius (°C)
time	t	seconds (s)
volume	V	metres cubed (m^3)
work done	W	joules (J)
kinetic energy	E_k	joules (J)
gravitational potential energy	E_p	joules (J)
elastic potential energy	E_e	joules (J)

The following equations need to be remembered for physics paper 1. Cover up the right-hand side and regularly test yourself to check that you can remember the equations.

Kinetic energy	kinetic energy = 0.5 × mass × (speed)^2
	$E_k = \frac{1}{2} m v^2$
Gravitational potential energy	g.p.e. = mass × gravitational field strength × height
	$E_p = m g h$
Power	power = $\frac{energy\ transferred}{time}$; $P = \frac{E}{t}$
	power = $\frac{work\ done}{time}$; $P = \frac{W}{t}$
Efficiency	efficiency = $\frac{useful\ output\ energy\ transfer}{total\ input\ energy\ transfer}$
	efficiency = $\frac{useful\ power\ output}{total\ power\ input}$
Charge flow	charge flow = current × time
	$Q = I t$
Current, resistance and potential difference	potential difference = current × resistance
	$V = I R$
Resistance in series	$R_{total} = R_1 + R_2$
Electrical power	power = potential difference × current
	$P = V I$
	power = (current)^2 × resistance
	$P = I^2 R$
Energy transfers	energy transferred = power × time
	$E = P t$
	energy transferred = charge flow × potential difference
	$E = Q V$
Density	density = $\frac{mass}{volume}$
	$\rho = \frac{m}{V}$

These equations are given, but you still need to know how to use them. The equation in **bold** is physics only (not on the combined science course).

Elastic potential energy	elastic potential energy = 0.5 × spring constant × (extension)^2
	$E_e = \frac{1}{2} k e^2$
Specific heat capacity	change in thermal energy = mass × specific heat capacity × temperature change
	$\Delta E = m c \Delta \theta$
Specific latent heat	thermal energy for a change of state = mass × specific latent heat
	$E = m L$
Pressure in gases	**pressure × volume = constant**
	$p V = constant$

Kinetic energy (combined science)

Any object that is moving has a store of kinetic energy.

The equation for kinetic energy is: $E_k = \frac{1}{2} m v^2$

You need to remember this equation!

Basic questions: No rearranging or unit conversion needed.

Model example: A football of mass 0.5 kg is kicked to a velocity of 12 m/s. Calculate the amount of kinetic energy stored by the football. [3 marks]

Write down equation:

$$E_k = \frac{1}{2} m v^2$$

Substitute variables into equation:

$$E_k = \frac{1}{2} \times 0.5 \times 12^2$$ [1 mark]

Calculate answer:

$$E_k = \frac{1}{2} \times 0.5 \times 12^2 = 36$$ [1 mark]

Write units:

joules (J) [1 mark]

Q1. A cat of mass 4 kg is travelling at the following velocities. Calculate the amount of kinetic energy stored by the cat at each speed:

a) 10 m/s b) 2 m/s c) 0.5 m/s d) 1 m/s [3 marks each]

a) ....

b) ....

c) ....

d) ....

Q2. A motorbike of mass 200 kg is travelling at the following velocities. Calculate the amount of kinetic energy stored by the motorbike at each speed:

a) 10 m/s b) 40 m/s c) 5 m/s d) 2 m/s [3 marks each]

a) ....

b) ....

c) ....

d) ....

Medium questions: Rearranging and unit conversion needed (hints in boxes).

Q3. A tennis player hits a tennis ball that has a mass of 56 g. Calculate the velocity of the tennis ball for each kinetic energy:

$g \rightarrow kg \div 1000$

a) 0.28 J b) 4.48 J c) 0.07 J d) 0.63 J [4 marks each]

a) ....

b) ....

c) ....

d) ....

Q4. A meteor shower can be seen in the night sky. If all of the meteors have a velocity of 12 000 m/s, calculate the mass of the meteor for each kinetic energy:

$$MJ \rightarrow J \times 1\,000\,000$$

| **a)** 180 MJ | **b)** 72 MJ | **c)** 108 MJ | **d)** 288 MJ | [4 marks each] |

a) ....

b) ....

c) ....

d) ....

Hard questions: Rearranging, use of standard form and unit conversion needed.

Q5. A racing car is travelling with a kinetic energy store of 4.8×10^6 J and has a mass of 1500 kg. Calculate the velocity of the racing car. [3 marks]

....

....

Q6. A motorbike is travelling at a velocity of 50 m/s and has a kinetic energy store of 225 kJ. Calculate the mass of the motorbike. [4 marks]

....

....

Q7. A bullet of mass 8 g travels at a velocity of 1500 km/h. Calculate the kinetic energy store of the bullet. [4 marks]

....

....

Q8. A cricketer bowls a cricket ball of mass of 160 g and gives the ball a kinetic energy store of 128 J. Calculate the velocity of the cricket ball. [4 marks]

....

....

Q9. A child is travelling on a skateboard at a velocity of 5 m/s. The child and skateboard have a combined kinetic energy store of 500 J and the skateboard alone has a mass of 2 kg. Calculate the mass of the child. [4 marks]

....

....

[Total marks   / 75]

Elastic potential energy (combined science)

Any object that has been stretched or compressed stores elastic potential energy.

The equation for elastic potential energy is: $E_e = \frac{1}{2} k e^2$

This equation is given to you!

Basic questions: No rearranging or unit conversion needed.

Model example: A spring has a spring constant of 50 N/m. Calculate the elastic potential energy stored by the spring if it is extended by a distance of 0.2 m.　　　[3 marks]

Write down equation:　　　　　　　　$E_e = \frac{1}{2} k e^2$

Substitute variables into equation:　　　$E_e = \frac{1}{2} \times 50 \times 0.2^2$　　　[1 mark]

Calculate answer:　　　　　　　　　$E_e = \frac{1}{2} \times 50 \times 0.2^2 = 1$　　　[1 mark]

Write units:　　　　　　　　　　　joules (J)　　　[1 mark]

Q1. A spring has a spring constant of 40 N/m. Calculate the elastic potential energy stored by the spring for each extension:

a) 2 m　　　　**b)** 1.5 m　　　　**c)** 0.3 m　　　　**d)** 0.1 m　　　[3 marks each]

a) ....

b) ....

c) ....

d) ....

Q2. An elastic band has a spring constant of 18 N/m. Calculate the elastic potential energy stored for each extension:

a) 0.2 m　　　　**b)** 0.05 m　　　　**c)** 1.1 m　　　　**d)** 3 m　　　[3 marks each]

a) ....

b) ....

c) ....

d) ....

Medium questions: Rearranging needed. Some unit conversion needed (hints in boxes).

Q3. A spring has a spring constant of 400 N/m. Calculate the extension for each of the stored potential energies:

kJ → J × 1000

a) 200 J　　　　**b)** 0.8 kJ　　　　**c)** 50 J　　　　**d)** 1.8 kJ　　　[3–4 marks each]

a) ....

b) ....

c) ....

d) ....

Q4. A spring stores 40 J of elastic potential energy. Calculate the spring constant of the spring for each of the extensions:

> cm → m ÷ 100
> mm → m ÷ 1000

| **a)** 50 cm | **b)** 80 mm | **c)** 160 mm | **d)** 4 cm | [4 marks each] |

a) ....

b) ....

c) ....

d) ....

Hard questions: Word-based questions that involve rearranging and use of standard form.

Q5. Each arm of a car's suspension has coil springs of spring constant 4×10^4 N/m. When the car goes over a speed bump, each coil spring stores 2 J in elastic potential energy. Calculate how much the coil spring is compressed by. [3 marks]

....

....

Q6. At maximum extension, a bungee cord stores 1.5×10^5 J of elastic potential energy. If a bungee jumper extends the cord by a maximum of 50 m, calculate the spring constant. [3 marks]

....

....

Mixed practice: Requires more than one equation.

Q7. An arrow of mass 250 g is fired by a bow. The string on the bow has a spring constant of 1500 N/m, and it is extended by 10 cm.

a) Calculate the maximum elastic potential energy stored by the bow. [4 marks]

....

....

b) Assuming that all the elastic potential energy store is converted into the kinetic energy store, calculate the initial velocity of the arrow. [4 marks]

....

....

[Total marks   / 68]

Gravitational potential energy (combined science)

Objects that are raised a height above the ground store gravitational potential energy. The gravitational field strength on Earth is equal to 9.8 N/kg.

The equation for gravitational potential energy is: $E_p = mgh$

You need to remember this equation!

Basic questions: No rearranging or unit conversion needed.

Model example: A phone of mass 0.12 kg is raised to a height of 1.8 m. Calculate the gravitational potential energy store of the phone. [3 marks]

Write down equation:	$E_p = mgh$	
Substitute variables into equation:	$E_p = 0.12 \times 9.8 \times 1.8$	[1 mark]
Calculate answer:	$E_p = 0.12 \times 9.8 \times 1.8 = 2.1$	[1 mark]
Write units:	joules (J)	[1 mark]

Q1. A book of mass 0.4 kg is put on a shelf. Calculate the gravitational potential energy store of the book at each shelf height:

a) 1 m   **b)** 0.8 m   **c)** 1.5 m   **d)** 2.5 m   [3 marks each]

a) ....

b) ....

c) ....

d) ....

Medium questions: Rearranging needed.

Q2. A student is walking up a hill to get to school. The mass of the student is 50 kg. Calculate the height of the hill climbed for each gravitational potential energy store:

a) 4900 J   **b)** 980 J   **c)** 3920 J   **d)** 5880 J   [3 marks each]

a) ....

b) ....

c) ....

d) ....

Q3. Different balls are dropped from a height of 5 m. Calculate the mass of the ball for each gravitational potential energy store:

a) 4.9 J   **b)** 24.5 J   **c)** 19.6 J   **d)** 34.3 J   [3 marks each]

a) ....

b) ....

c) ....

d) ....

Hard questions: Rearranging, use of standard form and unit conversion needed.

Q4. A builder is working on the roof of a block of flats. The builder has a mass of 70 kg and a gravitational potential energy store of 2.1×10^4 J. Calculate the height of the block of flats. [3 marks]

....

....

Q5. A cat is walking along the top of a fence. The cat has a mass of 4 kg and has a gravitational potential energy store of 78.4 J. Calculate the height of the fence. [3 marks]

....

....

Q6. An aeroplane is flying at a height of 8 km and has a gravitational potential energy store of 3.2×10^9 J. Calculate the mass of the aeroplane. [4 marks]

....

....

Mixed practice: Requires more than one equation.

Q7. A roof tile has become loose and falls to the ground. The roof tile has a mass of 800 g and the height of the roof is 8 m.

a) Calculate the gravitational potential energy store of the roof tile. [4 marks]

....

....

b) Assuming that all the gravitational potential energy store is converted into kinetic energy, calculate the velocity of the tile when it reaches the ground. [3 marks]

....

....

Q8. A tennis ball of mass 56 g is dropped from a height of 6 m. Calculate the velocity of the tennis ball when it reaches the ground. [6 marks]

....

....

....

[Total marks   / 59]

Specific heat capacity (combined science)

The specific heat capacity of a material is the amount of energy needed to heat up one kilogram of the material by a temperature of one degree Celsius.

This equation is given to you!

The equation to calculate change in thermal energy is: $\Delta E = m c \Delta \theta$

Basic questions: No rearranging or unit conversion needed.

Model example: Oil has a specific heat capacity of 540 J/kg °C. Calculate the amount of thermal energy that is needed to heat 2 kg of oil by a temperature of 10 °C. [3 marks]

Write down equation:	$\Delta E = m c \Delta \theta$	
Substitute variables into equation:	$\Delta E = 2 \times 540 \times 10$	[1 mark]
Calculate answer:	$\Delta E = 2 \times 540 \times 10 = 11\,000$	[1 mark]
Write units:	joules (J)	[1 mark]

Q1. The specific heat capacity of water is 4200 J/kg °C. Calculate the amount of thermal energy that is needed to raise the temperature of 0.4 kg of water by each temperature:

a) 10 °C   **b)** 25 °C   **c)** 60 °C   **d)** 2 °C   [3 marks each]

a) ....

b) ....

c) ....

d) ....

Q2. The specific heat capacity of iron is 450 J/kg °C. Calculate the amount of thermal energy that is needed to raise the temperature of 2 kg of iron by each temperature:

a) 1 °C   **b)** 2.5 °C   **c)** 400 °C   **d)** 1200 °C   [3 marks each]

a) ....

b) ....

c) ....

d) ....

Q3. Some water has just been boiled in a kettle and is left to cool from an initial temperature of 100 °C to a temperature of 50 °C. There is 0.8 kg of water in the kettle and the specific heat capacity of water is 4200 J/kg °C. Calculate how much the thermal energy store of the surroundings has been raised by. [3 marks]

....

....

Medium questions: Rearranging needed.

Q4. Copper has a specific heat capacity of 390 J/kg °C. If the thermal energy store of a copper block is increased by 8000 J, calculate the increase in temperature of the block for each mass:

a) 0.1 kg   **b)** 0.8 kg   **c)** 20 kg   **d)** 25 kg   [3 marks each]

a) ....

b) ....

c) ....

d) ....

Q5. The oil inside an oil radiator has a specific heat capacity of 540 J/kg °C. If the thermal energy store of the oil is increased by 6000 J, calculate the mass of oil for each increase in temperature:

a) 20 °C   **b)** 30 °C   **c)** 12 °C   **d)** 8 °C   [3 marks each]

a) ....

b) ....

c) ....

d) ....

Hard questions: Rearranging and use of standard form needed.

Q6. A bath is filled with water and the water is given 2×10^7 J of thermal energy. This leads to an increase of 40 °C in the temperature of the water. Calculate the mass of the water in the bath. The specific heat capacity of water is 4200 J/kg °C.   [3 marks]

....

....

Q7. The Sun heats 8×10^5 kg of sand on a beach and raises its thermal energy store by 2.2×10^9 J. Calculate the increase in temperature of the sand. The specific heat capacity of sand is 830 J/kg °C.   [3 marks]

....

....

Q8. An 800 g aluminium block is heated and gains 1.5×10^4 J of energy into its thermal energy store. The aluminium block is at an initial temperature of 20 °C, calculate the temperature of the block after it has been heated. The specific heat capacity of aluminium is 900 J/kg °C.   [4 marks]

....

....

[Total marks   / 61]

Power (combined science)

Power is the rate of energy transfer. Another phrase for energy transfer is work done.

The equation for power (in terms of energy transferred) is: $P = \frac{E}{t}$

The equation for power (in terms of work done) is: $P = \frac{W}{t}$

You need to remember these equations!

Basic questions: No rearranging or unit conversion needed.

Model example: A toaster uses 150 000 J of energy in a time of 75 seconds. Calculate the power of the toaster. [3 marks]

Write down equation:

$$P = \frac{E}{t}$$

Substitute variables into equation:

$$P = \frac{150\,000}{75}$$ [1 mark]

Calculate answer:

$$P = \frac{150\,000}{75} = 2000$$ [1 mark]

Write units:

watts (W) [1 mark]

Q1. A hairdryer transfers 100 000 J of energy. Calculate the power of the hairdryer if it is on for the following times:

a) 50 s   b) 40 s   c) 100 s   d) 110 s   [3 marks each]

a) ....

b) ....

c) ....

d) ....

Q2. A phone charger transfers 200 J of energy. Calculate the power of the phone charger if it is on for the following times:

a) 40 s   b) 16 s   c) 5 s   d) 12.5 s   [3 marks each]

a) ....

b) ....

c) ....

d) ....

Q3. A laptop transfers 16 000 J of energy in a time of 200 seconds. Calculate the power of the laptop. [3 marks]

....

....

Medium questions: Rearranging and unit conversion needed (hints in box).

Q4. A games console has a power of 120 W. Calculate how long the console has been on for each energy transfer:

a) 60 000 J   **b)** 240 000 J   **c)** 48 000 J   **d)** 1 200 000 J   [3 marks each]

a) ....

b) ....

c) ....

d) ....

Q5. An electric radiator has a power of 3 kW. Calculate the amount of energy transferred by the radiator if it is used for the following times:

kW → W × 1000
minutes → s × 60

a) 2 minutes   **b)** 5 minutes   **c)** 18 minutes   **d)** 60 minutes   [5 marks each]

a) ....

b) ....

c) ....

d) ....

Hard questions: Rearranging and unit conversion needed.

Q6. An electric oven has a power of 2.4 kW. The oven has been used for a time of 1 hour. Calculate how much energy has been transferred by the oven.   [5 marks]

....

....

Q7. A microwave of power 0.8 kW heats a ready meal for a time of 4 minutes. Calculate how much energy the microwave has transferred.   [5 marks]

....

....

Mixed practice: Requires more than one equation.

Q8. An electric motor raises a lift of total mass 500 kg by a height of 20 m in a time of 30 seconds. Calculate the power of the electric motor. Gravitational field strength = 9.8 N/kg.   [4 marks]

....

....

[Total marks   / 73]

Efficiency (combined science)

The more energy or power a device wastes, the less efficient it is. Efficiency is a measure of the proportion of energy or power that is used usefully.

The equation for efficiency (in terms of energy transfer) is: $efficiency = \frac{useful\ output\ energy\ transfer}{total\ input\ energy\ transfer}$

The equation for efficiency (in terms of power) is: $efficiency = \frac{useful\ power\ output}{total\ power\ input}$

You need to remember these equations!

Basic questions: No rearranging or unit conversion needed.

Model example: An electrical device has a total input energy transfer of 1200 J and a useful output energy transfer of 900 J. Calculate the efficiency of the device. [2 marks]

Write down equation: $efficiency = \frac{useful\ output\ energy\ transfer}{total\ input\ energy\ transfer}$

Substitute variables into equation: $efficiency = \frac{900}{1200}$ [1 mark]

Calculate answer: $efficiency = \frac{900}{1200} = 0.75$ [1 mark]

Q1. An electrical device has a total input energy transfer of 4000 J. Calculate the efficiency for each useful output energy transfer:

a) 3000 J   **b)** 2000 J   **c)** 3600 J   **d)** 3800 J   [2 marks each]

a) ....

b) ....

c) ....

d) ....

Q2. An electrical device has a useful power output of 500 W. Calculate the efficiency for each total power input:

a) 1000 W   **b)** 550 W   **c)** 5000 W   **d)** 1500 W   [2 marks each]

a) ....

b) ....

c) ....

d) ....

Medium questions: Rearranging needed.

Q3. A car has a total input energy of 200 000 J from the chemical energy store of its petrol. Calculate the output kinetic energy for each efficiency:

a) 0.6   **b)** 0.8   **c)** 0.15   **d)** 0.55   [3 marks each]

a) ....

b) ....

c) ....

d) ....

Q4. A transformer has a useful output power of 50 000 W. Calculate the total power input for each efficiency:

    a) 0.5   **b)** 0.2   **c)** 0.25   **d)** 0.4   [3 marks each]

a) ....

b) ....

c) ....

d) ....

Hard questions: Rearranging and unit conversion needed.

Q5. An electrical device has an efficiency of only 0.005. If there is a total power input to the device of 2 kW, calculate the useful power output. Give your answer in watts. [3 marks]

....

....

Q6. A typical coal power plant has an efficiency of 0.4. If the power plant output power is of 5×10^8 W, calculate its total power input. Give your answer in megawatts. [3 marks]

....

....

Mixed practice: Requires more than one equation.

Q7. An electric motor on a crane lifts a crate of mass 200 kg a distance of 40 m. If there is a total energy input of 156.8 kJ, calculate the efficiency of the electric motor.
Gravitational field strength = 9.8 N/kg [5 marks]

....

....

Q8. A rubber band of spring constant 80 N/m is stretched by a distance of 10 cm. The rubber band is then released and has a kinetic energy store of 0.3 J. Calculate the efficiency of the energy transfer into this kinetic energy store. [5 marks]

....

....

[Total marks   / 56]

Charge flow (combined science)

Current is a measure of how much charge is flowing every second (it is the rate of flow of charge).

The equation that links charge, current and time is: $Q = It$

> **You need to remember this equation!**

Basic questions: No rearranging or unit conversion needed.

Model example: An electric circuit has a current of 0.5 A flowing through it for a time of 30 seconds. Calculate how much charge has flowed through the circuit. [3 marks]

Write down equation:	$Q = It$	
Substitute variables into equation:	$Q = 0.5 × 30$	[1 mark]
Calculate answer:	$Q = 0.5 × 30 = 15$	[1 mark]
Write units:	coulombs (C)	[1 mark]

Q1. A pair of hair straighteners have a current of 6 A passing through them. Calculate the charge that has flowed through the hair straighteners for each time of use:

a) 90 seconds   b) 150 seconds   c) 60 seconds   d) 400 seconds   [3 marks each]

a) ....

b) ....

c) ....

d) ....

Q2. A remote controlled car has a current of 0.5 A passing through it. Calculate the charge that has flowed through it for each time of use:

a) 500 seconds   b) 1800 seconds   c) 2800 seconds   d) 600 seconds   [3 marks each]

a) ....

b) ....

c) ....

d) ....

Medium questions: Rearranging and unit conversion needed (hints in boxes).

Q3. An electric radiator has a charge of 18 000 C flowing through it. Calculate the current flowing through the radiator if it is on for the following times:

> minutes → s × 60
> hours → s × 3600

a) 30 minutes   b) 75 minutes   c) 1 hour   d) 4 hours   [4 marks each]

a) ....

b) ....

c) ....

d) ....

Q4. A printer has a charge of 90 C flowing through it. Calculate the current flowing for each of the times:

a) 1 minute   **b)** 3 minutes   **c)** 4 minutes   **d)** 7 minutes   [4 marks each]

a) ....

b) ....

c) ....

d) ....

Q5. A laptop charger has a current of 2 A flowing through it. For each total charge flow, calculate how long the laptop has been charging for:

$$kC \rightarrow C \times 1000$$

a) 2 kC   **b)** 1.5 kC   **c)** 5 kC   **d)** 8.4 kC   [4 marks each]

a) ....

b) ....

c) ....

d) ....

Hard questions: Rearranging, use of standard form and unit conversion needed.

Q6. **a)** Electrons are the charge carriers in metals. If one electron flows past a point every 5 seconds, there is a current of 3.2×10^{-20} C. Calculate the charge on an electron.   [3 marks]

....

....

b) Using your answer to part a), calculate the current that would flow if 5×10^{16} electrons pass a point in a time of 2 ms.   [4 marks]

....

....

c) Using your answer to part a), calculate the current that would flow if 2×10^{10} electrons pass a point in a time of 6 ms.   [4 marks]

....

....

[Total marks   / 83]

Potential difference, current and resistance (combined science)

Ohm's law tells us that the current flowing through a resistor is directly proportional to the potential difference across it (as long as the temperature of the resistor is constant).

The equation that links potential difference, current and resistance is: $V = IR$

You need to remember this equation!

Basic questions: No rearranging or unit conversion needed.

Model example: A thermistor of resistance 500 Ω is in an electric circuit. If there is a current of 0.2 A flowing through the thermistor, calculate the potential difference across it. [3 marks]

Write down equation:	$V = IR$	
Substitute variables into equation:	$V = 0.2 \times 500$	[1 mark]
Calculate answer:	$V = 0.2 \times 500 = 100$	[1 mark]
Write units:	volts (V)	[1 mark]

Q1. A resistor of resistance 20 Ω is in an electric circuit. Calculate the potential difference across the resistor for each current:

    a) 0.5 A   **b)** 0.8 A   **c)** 4 A   **d)** 15 A   [3 marks each]

    a) ....

    b) ....

    c) ....

    d) ....

Q2. A lamp has a resistance of 50 Ω. Calculate the potential difference across the lamp for each current:

    a) 2 A   **b)** 0.4 A   **c)** 0.1 A   **d)** 1.5 A   [3 marks each]

    a) ....

    b) ....

    c) ....

    d) ....

Medium questions: Rearranging and unit conversion needed (hints in boxes).

Q3. A thermistor has a potential difference of 12 V across it. Calculate the resistance of the thermistor for each current flowing through it:

mA → A ÷ 1000

    a) 200 mA   **b)** 50 mA   **c)** 4 mA   **d)** 30 mA   [4 marks each]

a) ....

b) ....

c) ....

d) ....

Q4. A light dependent resistor (LDR) has a potential difference of 60 V across it. Calculate the current flowing through it for each of the resistances:

$k\Omega \rightarrow \Omega \times 1000$

a) 1 kΩ   b) 5 kΩ   c) 3 kΩ   d) 60 kΩ   [4 marks each]

a) ....

b) ....

c) ....

d) ....

Hard questions: Rearranging and unit conversion needed.

Q5. A high voltage power supply places a potential difference of 400 kV across a circuit of resistance 800 Ω. Calculate the current that is flowing through the circuit.   [4 marks]

....

....

Q6. A Van de Graaff generator produces a potential difference of 5 MV. Calculate the current produced if this was discharged into a circuit of resistance 25 kΩ.   [5 marks]

....

....

Mixed practice: Requires more than one equation.

Q7. A charge of 130 C flows in a circuit in a time of 10 seconds. If there is a resistor of 5 Ω resistance in the circuit, calculate the potential difference across the resistor.   [5 marks]

....

....

Q8. There is a potential difference of 12 V across a lamp. If the lamp has a resistance of 60 Ω, calculate how much charge flows through the lamp in a time of 2 minutes.   [6 marks]

....

....

[Total marks   / 76]

Series and parallel circuits (combined science)

In a series circuit:
- The current is the same through all electrical components.
- The supply potential difference is shared across components.
- The resistance of two electrical components add together as: $R_{total} = R_1 + R_2$

In a parallel circuit:
- The total current through the whole circuit is shared across different paths.
- The potential difference is the same across each path.
- The total resistance of two electrical components is less than either individual resistance.

> **You need to remember this equation!**

Basic questions: Calculating total resistance in a series circuit.

Q1. Two resistors are in a series circuit. Calculate the total resistance of each pair of resistances:

a) $R_1 = 10\ \Omega$, $R_2 = 5\ \Omega$ **b)** $R_1 = 200\ \Omega$, $R_2 = 25\ \Omega$

c) $R_1 = 400\ \Omega$, $R_2 = 2000\ \Omega$ **d)** $R_1 = 0.2\ \Omega$, $R_2 = 0.3\ \Omega$ [1 mark each]

a) ....

b) ....

c) ....

d) ....

Medium questions: Calculating current and potential difference in series circuits.

Model example: A battery supplies a potential difference of 9 V to a series circuit that contains a 10 Ω resistor and an 8 Ω resistor. Calculate the current that is flowing in the circuit.   [4 marks]

Calculate total resistance of circuit:	$R_{total} = 10 + 8 = 18\ \Omega$	[1 mark]
Substitute variables into $V = IR$:	$9 = I \times 18$	[1 mark]
Rearrange equation to make I the subject:	$I = \frac{9}{18} = 0.5$	[1 mark]
Write units:	amps (A)	[1 mark]

Q2. A power supply has a potential difference of 12 V across it and is connected in series with two resistors. One of the resistors has a resistance of 20 Ω and the other has a resistance of 40 Ω.

a) Calculate the total resistance of the series circuit.   [1 mark]

....

b) Calculate the current in the series circuit.   [3 marks]

....

c) Describe how the current varies at different positions in the circuit.   [1 mark]

....

d) Calculate the potential difference across the 20 Ω resistor. [3 marks]

....

e) Calculate the potential difference across the 40 Ω resistor. [1 mark]

....

f) An additional parallel branch to the circuit is added. State the potential difference across this branch of the circuit. [1 mark]

....

Hard questions: More complicated questions on a parallel circuit.

Q3. A power supply has a potential difference of 24 V across it and is connected to a parallel circuit with two paths.

The first path consists of a 20 Ω resistor and a 28 Ω resistor.

The second path consists of a 2 Ω resistor and a 14 Ω resistor.

a) State the potential difference across the first path. [1 mark]

....

b) Calculate the current flowing through the first path. [3 marks]

....

c) Calculate the potential difference across the 20 Ω resistor. [3 marks]

....

d) State the potential difference across the second path. [1 mark]

....

e) Calculate the current flowing through the second path. [3 marks]

....

f) Calculate the potential difference across the 14 Ω resistor. [3 marks]

....

g) Calculate the current through the whole circuit. [1 mark]

....

[Total marks   / 29]

Electrical power (combined science)

Power is the rate of energy transfer. The more energy that a circuit uses each second, the higher the power of the circuit.

You need to remember these equations!

The equation that links power, potential difference and current is: $P = VI$

The equation that links power, current and resistance is: $P = I^2 R$

Basic questions: No rearranging or unit conversion needed.

Model example: A laptop has a potential difference supply of 12 V and a current of 0.5 A flowing through it. Calculate the power of the laptop. [3 marks]

Write down equation: $P = VI$

Substitute variables into equation: $P = 12 \times 0.5$ [1 mark]

Calculate answer: $P = 12 \times 0.5 = 6$ [1 mark]

Write units: watts (W) [1 mark]

Q1. A light emitting diode (LED) has a potential difference of 6 V across it. Calculate the power of the LED for each current flowing through it:

    a) 2 A   **b)** 2.5 A   **c)** 0.8 A   **d)** 3 A   [3 marks each]

    a) ....

    b) ....

    c) ....

    d) ....

Q2. A lamp has a current of 2.5 A flowing through it. Calculate the power of the lamp for each resistance:

    a) 20 Ω   **b)** 5 Ω   **c)** 100 Ω   **d)** 4 Ω   [3 marks each]

    a) ....

    b) ....

    c) ....

    d) ....

Medium questions: Rearranging and unit conversion needed (hints in boxes).

Q3. An electric motor uses a power of 500 W. Calculate the current flowing through it for each of the resistances:

    $kΩ \rightarrow Ω \times 1000$

    a) 0.5 kΩ   **b)** 4 kΩ   **c)** 1.5 kΩ   **d)** 0.2 kΩ   [4 marks each]

a) ....

b) ....

c) ....

d) ....

Q4. A resistor has a power of 5 W. Calculate the potential difference across it for each of the currents flowing through it:

> mA → A ÷ 1000

    a) 250 mA   **b)** 80 mA   **c)** 400 mA   **d)** 750 mA   [4 marks each]

a) ....

b) ....

c) ....

d) ....

Hard questions: Rearranging and unit conversion needed.

Q5. A hairdryer transfers 2 kW of power and it is connected to mains electricity with a potential difference supply of 230 V. Calculate the current flowing through the hairdryer.   [4 marks]

....

....

Q6. A set of speakers has a power of 50 W and a current of 500 mA flowing into it. Calculate the total resistance of the speakers.   [4 marks]

....

....

Mixed practice: Requires more than one equation.

Q7. A lamp has a potential difference of 5 V across it and a resistance of 25 Ω. Calculate the power of the lamp.   [5 marks]

....

....

Q8. An electric motor does 50 kJ of work in 2 minutes. If the electric motor is supplied with a potential difference of 230 V, what is the current flowing through the motor?   [6 marks]

....

....

[Total marks   / 75]

Energy transfer (combined science)

Potential difference is a measure of how much work (transfer of energy) each coulomb of charge can do.

The equation that links energy transfer, charge flow and potential difference is: $E = Q V$

You need to remember this equation!

Basic questions: No rearranging or unit conversion needed.

Model example: A microwave is supplied with a potential difference of 230 V and there is an overall charge flow of 20 000 C. Calculate the energy transferred to the microwave. [3 marks]

Write down equation:	$E = Q V$	
Substitute variables into equation:	$E = 20\ 000 \times 230$	[1 mark]
Calculate answer:	$E = 20\ 000 \times 230 = 4\ 600\ 000$	[1 mark]
Write units:	joules (J)	[1 mark]

Q1. A charge of 4000 C flows through a resistor. Calculate the energy transferred to the resistor for each potential difference across it:

a) 6 V   **b)** 1.5 V   **c)** 20 V   **d)** 45 V   [3 marks each]

a) ....

b) ....

c) ....

d) ....

Q2. A television has a potential difference of 230 V supplied to it. Calculate the energy transferred to the TV for each charge flow into it:

a) 40 C   **b)** 500 C   **c)** 600 C   **d)** 3 C   [3 marks each]

a) ....

b) ....

c) ....

d) ....

Medium questions: Rearranging and unit conversion needed (hints in boxes).

Q3. A washing machine has a potential difference of 230 V supplied to it. Calculate the charge flow into the washing machine for each energy transfer to it:

MJ → J × 1 000 000

a) 4.6 MJ   **b)** 3.9 MJ   **c)** 1.2 MJ   **d)** 0.23 MJ   [4 marks each]

a) ....

b) ....

c) ....

d) ....

Q4. A torch has a potential difference of 3 V supplied to it. Calculate the charge flow into the torch for each energy transfer to it:

kJ → J × 1000

a) 0.15 kJ   **b)** 6 kJ   **c)** 7.5 kJ   **d)** 0.42 kJ   [4 marks each]

a) ....

b) ....

c) ....

d) ....

Hard questions: Rearranging and unit conversion needed.

Q5. An electric oven has 6.9 MJ of energy transferred to it. The oven is supplied with a potential difference of 230 V. Calculate how much charge has flowed into it.   [4 marks]

....

....

Q6. A phone has a total energy of 144 kJ transferred to it. If 12 000 C of charge flows into the phone, calculate the potential difference that is supplied to it.   [4 marks]

....

....

Mixed practice: Requires more than one equation.

Q7. A laptop is charged with a circuit that has a current of 0.2 A and a total resistance of 60 Ω. If a charge of 500 kC flows into the laptop, calculate the overall energy transferred to it.   [6 marks]

....

....

Q8. A light fitting has a current of 0.5 A flowing through it for a time of 5 minutes. If it is supplied with a potential difference of 230 V, calculate the energy transferred to the light fitting.   [6 marks]

....

....

[Total marks   / 76]

Density (combined science)

The density of an object is a measure of how much mass there is per unit volume (how many kilograms there are in each metre cubed of the material).

The equation that links density, mass and volume is: $\rho = \frac{m}{V}$

ρ (rho) is the symbol you use for density.

You need to remember this equation!

Basic questions: No rearranging or unit conversion needed.

Model example: A bath is filled with 0.2 m³ of water. This volume has a mass of 200 kg. Calculate the density of water.   [3 marks]

Write down equation: $\rho = \frac{m}{V}$

Substitute variables into equation: $\rho = \frac{200}{0.2}$ [1 mark]

Calculate answer: $\rho = \frac{200}{0.2} = 1000$ [1 mark]

Write units: kilograms per metre cubed (kg/m³)   [1 mark]

Q1. A material has a mass of 800 kg. Calculate its density for each of the volumes:

    a) 0.5 m³   **b)** 2 m³   **c)** 0.8 m³   **d)** 1.6 m³   [3 marks each]

    a) ....

    b) ....

    c) ....

    d) ....

Q2. A material has a mass of 2 kg. Calculate its density for each of the volumes:

    a) 0.02 m³   **b)** 0.1 m³   **c)** 0.4 m³   **d)** 0.5 m³   [3 marks each]

    a) ....

    b) ....

    c) ....

    d) ....

Medium questions: Rearranging and unit conversion needed (hints in boxes).

Q3. A metal has a density of 12 000 kg/m³. Calculate the volume of the metal for each of the masses:

g → kg ÷ 1000

    a) 100 g   **b)** 800 g   **c)** 500 g   **d)** 40 g   [4 marks each]

a) ....

b) ....

c) ....

d) ....

Q4. Water has a density of 1000 kg/m^3. Calculate the mass of each of the volumes of water

$cm^3 \rightarrow m^3 \div 1\,000\,000$

a) 300 cm^3 **b)** 1500 cm^3 **c)** 4000 cm^3 **d)** 800 cm^3 [4 marks each]

a) ....

b) ....

c) ....

d) ....

Hard questions: Rearranging or unit conversion needed.

Q5. Water has a density of 1000 kg/m^3. A typical water bottle has a mass of 330 g. Calculate the volume of water inside one of these bottles. [4 marks]

....

....

Q6. A rover on Mars takes a soil sample of mass 5 µg. The density of the sample is 1600 kg/m^3. Calculate the volume of the sample. [4 marks]

....

....

Q7. Archimedes is tasked by the King of Syracuse to find out whether or not his crown is made of pure gold. The density of Gold is 19 300 kg/m^3. The mass of the crown is 800 g and its volume is 52 cm^3. Determine whether or not the crown is made of pure gold. [6 marks]

....

....

Mixed practice: Requires more than one equation.

Q8. A pallet of bricks is being lifted by a crane to a height of 20 m above the ground. At this height, they have a gravitational potential energy store of 784 kJ. If the bricks have a volume of 2 m^3, calculate the density of the bricks. Gravitational field strength = 9.8 N/kg. [6 marks]

....

....

[Total marks   / 76]

Specific latent heat (combined science)

The specific latent heat of a material is the energy required to change the state of 1 kg of that material (without a change in temperature).

The equation to calculate the energy for a change of state is: $E = m L$

> This equation is given to you!

Basic questions: No rearranging or unit conversion needed.

Model example: The specific latent heat of fusion of water is 334 000 J/kg. Calculate how much energy is needed to melt 0.2 kg of ice. [3 marks]

Write down equation: $E = m L$

Substitute variables into equation: $E = 0.2 \times 334\,000$ [1 mark]

Calculate answer: $E = 0.2 \times 334\,000 = 67\,000$ [1 mark]

Write units: joules (J) [1 mark]

Q1. The specific latent heat of vaporisation of oxygen is 213 000 J/kg. Calculate how much energy is needed to evaporate each mass of oxygen:

a) 0.02 kg   b) 0.1 kg   c) 0.08 kg   d) 0.22 kg   [3 marks each]

a) ....

b) ....

c) ....

d) ....

Q2. The specific latent heat of fusion of oxygen is 13 900 J/kg. Calculate how much energy is needed to melt each mass of oxygen:

a) 0.5 kg   b) 0.2 kg   c) 0.8 kg   d) 0.04 kg   [3 marks each]

a) ....

b) ....

c) ....

d) ....

Medium questions: Rearranging needed.

Q3. Helium has a specific latent heat of vaporisation of 21 000 J/kg. Calculate the mass of helium that can be evaporated with the following energies:

a) 42 000 J   **b)** 10 500 J   **c)** 7000 J   **d)** 3500 J   [3 marks each]

a) ....

b) ....

c) ....

d) ....

Q4. Carbon dioxide has a specific latent heat of fusion of 184 000 J/kg. Calculate the mass of carbon dioxide that can be melted with the following energies:

a) 9200 J   **b)** 46 000 J   **c)** 368 000 J   **d)** 73 600 J   [3 marks each]

a) ....

b) ....

c) ....

d) ....

Hard questions: Rearranging and unit conversion needed.

Q5. It takes 400 kJ of energy to evaporate 2000 g of nitrogen. Calculate the specific latent heat of vaporisation of nitrogen. Give your answer in units of J/kg.   [4 marks]

....

....

Q6. It takes 143.6 kJ of energy to condense 250 g of carbon dioxide. Calculate the specific latent heat of vaporisation of carbon dioxide. Give your answer in units of J/kg.   [4 marks]

....

....

Q7. It takes 9.04 MJ of energy to evaporate 4000 g of water. Calculate the specific latent heat of vaporisation of water. Give your answer in units of J/kg.   [4 marks]

....

....

[Total marks   / 60]

Pressure in gases (physics only)

If a gas is compressed then the pressure of the gas increases. If a gas is expanded, then the pressure decreases.

| This equation is given to you! |

If the number of particles in a gas and the temperature of the gas are constant then:

$pV = constant$

Basic questions: No rearranging or unit conversion needed.

Model example: Air is inside a car tyre at a pressure of 220 000 Pa. If the volume of the tyre is 0.005 m^3, calculate the 'constant'.

[3 marks]

Write down equation: $pV = constant$

Substitute variables into equation: $constant = 220\ 000 \times 0.005$ [1 mark]

Calculate answer: $constant = 220\ 000 \times 0.005 = 1100$ [1 mark]

Write units: pascal metres cubed (Pa m^3) [1 mark]

Q1. A gas is held inside a canister of volume 0.02 m^3. The pressure inside the canister is changed by adding or removing gas molecules. Calculate the "constant" for each pressure:

a) 200 000 Pa **b)** 250 000 Pa **c)** 850 000 Pa **d)** 300 000 Pa [3 marks each]

a) ....

b) ....

c) ....

d) ....

Medium questions: Rearranging and use of standard form needed.

Model example: A gas is in a sealed container of initial pressure 400 000 Pa and volume 0.01 m^3. Calculate the pressure if the container is expanded to a volume of 0.04 m^3. Assume that the temperature remains constant.

[5 marks]

Write down equation: $pV = constant$

Substitute variables into equation: $constant = 400\ 000 \times 0.01$ [1 mark]

Calculate constant: $constant = 400\ 000 \times 0.01 = 4000$ Pa m^3 [1 mark]

The constant is the same even though the volume has changed. Substitute variables into equation:
$pV = constant$
$p \times 0.04 = 4000$ [1 mark]

Rearrange and calculate: $p = \frac{4000}{0.04} = 100\ 000$ [1 mark]

Write units: pascals (Pa) [1 mark]

Q2.	A balloon contains helium at an initial pressure of 120 000 Pa and a volume of 0.04 m^3. The balloon is compressed and the pressure increases. Calculate the volume for each pressure:

a) 125 000 Pa	**b)** 160 000 Pa	**c)** 140 000 Pa	**d)** 240 000 Pa	[5 marks each]

a) ....

b) ....

c) ....

d) ....

Q3.	A syringe contains a gas at an initial pressure of 102 000 Pa and volume of 10^{-5} m^3. The volume of the container is changed. Calculate the final pressure of the gas for each volume:

a) 5 × 10^{-6} m^3	**b)** 4 × 10^{-5} m^3	**c)** 8 × 10^{-6} m^3	**d)** 1.6 × 10^{-5} m^3	[5 marks each]

a) ....

b) ....

c) ....

d) ....

Hard questions: Rearranging and unit conversion needed.

Q4.	A gas of initial pressure of 800 000 Pa and volume of 200 cm^3 is released into a larger (and initially empty) container of volume 0.2 m^3. Calculate the pressure of the gas in the larger container. Assume that the temperature of the gas remains constant.	[6 marks]

....

....

....

Q5.	A car tyre is filled with air of pressure 250 kPa and has an initial volume of 0.006 m^3. It is compressed to a volume of 0.004 m^3 as the car goes over a speed bump. Calculate the pressure in the tyre at this compressed volume. Assume that the temperature of the air remains constant.	[6 marks]

....

....

....

[Total marks	/ 64]

Nuclear equations (combined science)

The nucleus of an atom can emit four types of radiation; an alpha particle, a beta particle, a gamma ray or a neutron.

You need to remember this!

An alpha particle is the same as a helium nucleus. When an alpha is emitted, the nucleus loses 2 protons and 2 neutrons. The mass number of the nucleus therefore goes down by 4 and the atomic number goes down by 2.

An example of an alpha decay is: $^{238}_{92}U \rightarrow \ ^{234}_{90}Th + \ ^{4}_{2}He$

A beta particle is the same as a fast moving electron. When a beta particle is emitted, a neutron in the nucleus is converted into a proton. The atomic number therefore goes up by 1.

An example of beta decay is: $^{225}_{88}Ra \rightarrow \ ^{225}_{89}Ac + \ ^{0}_{-1}e$

When a gamma ray is emitted, there is no change to the structure of the nucleus. The nucleus becomes more stable by transferring energy in a gamma ray.

When a neutron is emitted from the nucleus, the mass number decreases by 1 but the atomic number stays the same.

Basic questions: Identifying which type of radiation is emitted.

Q1. A $^{208}_{84}Po$ nucleus emits one type of nuclear radiation. State which type of radiation is emitted for each of the nuclei left behind:

a) $^{204}_{82}Pb$ **b)** $^{208}_{84}Po$ **c)** $^{208}_{85}At$ **d)** $^{207}_{84}Po$ [1 mark each]

a) ....

b) ....

c) ....

d) ....

Q2. A $^{131}_{53}I$ nucleus emits one type of nuclear radiation. State which type of radiation is emitted for each of the nuclei left behind:

a) $^{131}_{54}Xe$ **b)** $^{130}_{53}I$ **c)** $^{127}_{51}Sb$ **d)** $^{131}_{53}I$ [1 mark each]

a) ....

b) ....

c) ....

d) ....

Medium questions: Completing partial nuclear equations.

Q3. Complete the following partial nuclear equations for alpha decay:

a) $^{241}_{95}Am \rightarrow \quad Np + \ ^{4}_{2}He$ **b)** $^{222}_{88}Ra \rightarrow \ ^{218}_{86}Rn + \quad He$

c) $^{218}Po \rightarrow \ _{82}Pb + \ ^{4}_{2}He$ **d)** $_{83}Bi \rightarrow \ ^{207}Tl + \ ^{4}_{2}He$ [2 marks each]

a) ....

b) ....

c) ....

d) ....

Q4. Complete the following partial nuclear equations for beta decay:

a) $^{6}_{2}\text{He} \rightarrow \text{Li} + ^{0}_{-1}\text{e}$

b) $^{14}_{6}\text{C} \rightarrow ^{14}_{7}\text{N} + \text{e}$

c) $^{209}\text{Pb} \rightarrow _{83}\text{Bi} + ^{0}_{-1}\text{e}$

d) $_{26}\text{Fe} \rightarrow ^{52}\text{Co} + ^{0}_{-1}\text{e}$ [2 marks each]

a) ....

b) ....

c) ....

d) ....

Hard questions: Writing nuclear equations for more than one decay.

Q5. A $^{238}_{92}\text{U}$ nucleus undergoes one alpha decay to form a Th nucleus and then undergoes one beta decay to form a Pa nucleus. Identify the atomic and mass numbers of the Pa nucleus.   [4 marks]

....

....

....

Q6. A $^{214}_{82}\text{Pb}$ nucleus undergoes a beta decay to form a Bi nucleus and then a second beta decay to form a Po nucleus. Identify the atomic and mass numbers of the Po nucleus.   [4 marks]

....

....

....

Q7. A $^{222}_{86}\text{Rn}$ nucleus undergoes an alpha decay to form a Po nucleus and then a second alpha decay to form a Pb nucleus. Identify the atomic and mass numbers of the Pb nucleus.   [4 marks]

....

....

....

[Total marks   / 36]

Half-lives (combined science)

Each radioactive material has a half-life. This is the time it takes for half of the radioactive nuclei to decay and emit nuclear radiation. Due to this, half-life has two definitions:

You need to remember this!

1. The time taken for half of the radioactive nuclei to decay.
2. The time taken for the radioactive count rate/activity to halve.

Basic questions: Count rate calculations.

Model example: An iodine-131 sample has an initial count rate of 1000 Bq. Calculate the count rate of the sample after 2 half-lives. [2 marks]

Halve initial count rate to get count rate after
1 half-life: $1000 \div 2 = 500$

Halve this again to get count rate after
2 half-lives: $500 \div 2 = 250$ [1 mark]

Write units: becquerel (Bq) [1 mark]

Q1. A sodium-24 sample has an initial count rate of 800 Bq. Calculate the count rate after each of the number of half-lives:

a) 1 half-life   b) 2 half-lives   c) 3 half-lives   d) 4 half-lives   [2 marks each]

a) ....

b) ....

c) ....

d) ....

Q2. A cobalt-60 sample has an initial count rate of 20 000 Bq. Cobalt-60 has a half-life of 5.3 years. Calculate the count rate after each of the times have passed:

a) 5.3 years   b) 10.6 years   c) 15.9 years   d) 21.2 years   [2 marks each]

a) ....

b) ....

c) ....

d) ....

Q3. A bone in a skeleton originally had 24 g of carbon-14. The half-life of carbon-14 is 5700 years. Calculate how old the skeleton is for each of the remaining masses of carbon-14. Give your answer to three significant figures.

a) 3 g   b) 6 g   c) 1.5 g   d) 12 g   [2 marks each]

a) ....

b) ....

c) ....

d) ....

Medium questions: Half-life calculations.

Q4. A sample of radium-226 has an initial activity of 4000 Bq. After 4800 years, the sample has an activity of 1000 Bq. Calculate the half-life of radium-226. [2 marks]

....

....

Q5. A sample of plutonium-229 has an initial activity of 600 Bq. After 360 s, the activity is 75 Bq. Calculate the half-life of plutonium-229. [2 marks]

....

....

Q6. A sample of curium-236 has an initial activity of 600 000 Bq. After a time of 900 s, the sample has an activity of 150 000 Bq. Calculate the half-life of curium-236. [2 marks]

....

....

Hard questions: Questions involving percentages or unit conversions.

Q7. Americium-230 has a half-life of 30 seconds. If a sample of this has an initial activity of 80 Bq, calculate the count rate after 2 minutes. [3 marks]

....

....

....

Q8. A radioactive sample has a half-life of 15 days. Calculate the time taken for the activity to go down to 12.5% of its initial value. [3 marks]

....

....

....

[Total marks / 36]

Physics (paper 2) summary sheet

The four topics in physics paper 2 are:

5. Forces
6. Waves
7. Magnetism and electromagnetism
8. Space physics (physics only)

Note that paper 1 content can be examined in paper 2. As with paper 1, it is vital to not only remember the necessary equations but also to remember the units. Use the table below to help with this memorisation. Cover the right-hand side and regularly quiz yourself to ensure all units are memorised.

Variables	Symbol	Units
acceleration	a	metres per second squared (m/s^2)
area	A	metres squared (m^2)
current	I	amps (A)
density	ρ	kilograms per metre cubed (kg/m^3)
distance	s	metres (m)
distance (to pivot)	d	metres (m)
extension	e	metres (m)
force	F	newtons (N)
frequency	f	hertz (Hz)
gravitational field strength	g	newtons per kilogram (N/kg)
length	l	metres (m)
magnetic flux density	B	tesla (T)
mass	m	kilograms (kg)
moment	M	newton metres (Nm)
momentum	p	kilograms metre per second (kg m/s)
number of turns	N	no units
potential difference	V	volts (V)
pressure	p	pascals (Pa)
speed	v	metres per second (m/s)
spring constant	k	newtons per metre (N/m)
time	t	seconds (s)
time period	T	seconds (s)
wavelength	λ	metres (m)
weight	W	newtons (N)
work done	W	joules (J)

The below equations need to be remembered. **Bold** equations are physics only.

Weight	weight = mass × gravitational field strength
	$W = m g$
Work done	work done = force × distance, $W = F s$
Forces and elasticity	force applied to a spring = spring constant × extension
(Hooke's law)	$F = k e$
Moments	**moment of a force = force × distance**
	$\boldsymbol{M = F d}$
Pressure	$\boldsymbol{pressure = \frac{force\ normal\ to\ a\ surface}{area\ of\ that\ surface}}; \boldsymbol{p = \frac{F}{A}}$
Speed	distance travelled = speed × time, $s = v t$
Acceleration	$acceleration = \frac{change\ in\ velocity}{time\ taken}; a = \frac{\Delta v}{t}$
Newton's second law	resultant force = mass × acceleration
	$F = m a$
Stopping distance	stopping distance = thinking distance + braking distance
Momentum	momentum = mass × velocity
	$p = m v$
Wave speed equation	wave speed = frequency × wavelength
	$v = f \lambda$

The below equations are given. Again, equations in **bold** are physics only.

Elastic potential energy	elastic potential energy = 0.5 × spring constant × (extension)^2
	$E_e = \frac{1}{2} k e^2$
Pressure in a fluid	**pressure due to a column of liquid =**
	height of column × density of liquid × gravitational field strength
	$\boldsymbol{p = h \rho g}$
Equation of motion	(final velocity)^2 – (initial velocity)^2 = 2 × acceleration × distance
	$v^2 - u^2 = 2 a s$
Changes in momentum	$\boldsymbol{force = \frac{change\ in\ momentum}{time\ taken}}; \boldsymbol{F = \frac{m \Delta v}{\Delta t}}$
Time period	$period = \frac{1}{frequency}; T = \frac{1}{f}$
Magnification	$\boldsymbol{magnification = \frac{image\ height}{object\ height}}$
Force on a current carrying wire	force on a conductor = magnetic flux density × current × length
	$F = B I l$
Transformers	$\frac{potential\ difference\ across\ primary\ coil}{potential\ difference\ across\ secondary\ coil} = \frac{number\ of\ turns\ in\ primary\ coil}{number\ of\ turns\ in\ secondary\ coil}$
	$\frac{V_p}{V_s} = \frac{n_p}{n_s}$
	potential difference across primary coil × current in primary coil =
	potential difference across secondary coil × current in secondary coil
	$V_p \times I_p = V_s \times I_s$

Weight (combined science)

Weight is an attractive force that acts on all masses in a gravitational field. On Earth, the gravitational field strength is equal to 9.8 N/kg.

The equation for weight is: $W = mg$

You need to remember this equation!

Basic questions: No rearranging or unit conversion needed.

Model example: A dog has a mass of 8 kg. Calculate the weight of the dog. [3 marks]

Write down equation: $W = mg$

Substitute variables into equation: $W = 8 \times 9.8$ [1 mark]

Calculate answer: $W = 8 \times 9.8 = 78$ [1 mark]

Write units: newtons (N) [1 mark]

Q1. Calculate the weight of each of these masses on Earth:

a) 2 kg   **b)** 0.5 kg   **c)** 400 kg   **d)** 15 kg   [3 marks each]

a) ....

b) ....

c) ....

d) ....

Q2. The gravitational field strength on the surface of the Moon is equal to 1.6 N/kg. Calculate the weight of the same masses on the Moon:

a) 2 kg   **b)** 0.5 kg   **c)** 400 kg   **d)** 15 kg   [3 marks each]

a) ....

b) ....

c) ....

d) ....

Q3. State how your answers between questions 1 and 2 differ. [1 mark]

....

Medium questions: Rearranging and unit conversion needed (hints in boxes).

Q4. A kitten is weighed (on Earth) at different ages. Calculate the weight of the kitten for each of the masses:

$g \rightarrow kg \div 1000$

a) 400 g   **b)** 600 g   **c)** 900 g   **d)** 1400 g   [4 marks each]

a) ....

b) ....

c) ....

d) ....

Q5. An exploration rover is launched onto different planets. The rover has a mass of 900 kg. Calculate the gravitational field strength for each of the weights:

kN → N × 1000

a) 3.33 kN   **b)** 7.92 kN   **c)** 8.82 kN   **d)** 1.44 kN   [4 marks each]

a) ....

b) ....

c) ....

d) ....

Hard questions: Rearranging and unit conversion needed.

Use g = 9.8 N/kg for all questions from this point.

Q6. A paracetamol tablet has a mass of 500 mg. Calculate the weight of the tablet. [4 marks]

....

....

Q7. A car has a weight of 12.9 kN. Calculate the mass of the car. [4 marks]

....

....

Mixed practice: Requires more than one equation.

Q8. A crane lifts a crate to a height of 40 m and transfers 78.4 kJ of energy into its gravitational potential energy store.

a) Calculate the weight of the crate. [6 marks]

....

....

b) An accident happens and the crate is dropped to the ground. Calculate the velocity of the crate when it reaches the ground. [3 marks]

....

....

[Total marks   / 74]

Resultant forces (combined science)

A force is an example of a vector – a quantity with both magnitude and direction.

Individual forces can be combined to form a resultant force. As force is a vector, the direction of the individual forces matters. Forces in the same direction are added, while forces in opposite directions are subtracted.

The resultant of forces can also be calculated by using scale drawings.

You need to remember this!

Basic questions: Forces in the same direction.

Model example: A crate is being pushed by two people. The first person exerts a force of 200 N to the right and the second person exerts a force of 300 N to the right. Calculate the resultant force of the people. [2 marks]

Add individual forces together: 200 + 300 = 500 N   [1 mark]

Write direction: To the right.   [1 mark]

Q1. Two people are both pushing a supermarket trolley to the right. The first person exerts a force of 50 N. Calculate the resultant force of the people if the second person exerts each of the forces:

a) 40 N   **b)** 50 N   **c)** 80 N   **d)** 10 N   [2 marks each]

a) ....

b) ....

c) ....

d) ....

Q2. Two tugboats are pulling a ship north. The first exerts a force of 20 000 N. Calculate the resultant force if the second tugboat exerts each of the forces:

a) 10 000 N   **b)** 5000 N   **c)** 20 000 N   **d)** 60 000 N   [2 marks each]

a) ....

b) ....

c) ....

d) ....

Medium questions: Forces in opposite directions.

Model example: A game of tug of war is being played. One team is pulling with a force of 5000 N to the right, while the other is pulling with a force of 4500 N to the left. What is the resultant force? [2 marks]

Subtract individual forces: 5000 – 4500 = 500 N   [1 mark]

Write direction: To the right.   [1 mark]

Q3. Two dogs are fighting over a bone. The first dog is pulling with a force of 400 N to the left. The second dog pulls to the right. Calculate the resultant force for each of the second forces:

a) 420 N   **b)** 300 N   **c)** 650 N   **d)** 860 N   [2 marks each]

a) ....

b) ....

c) ....

d) ....

Q4. A rocket has been launched upwards from the surface of Earth. The rocket has a constant upwards thrust force of 3.5×10^7 N. As the rocket goes further away from Earth, the gravitational field strength decreases. The mass of the rocket also decreases due to fuel being burnt. Calculate the resultant force on the rocket for each weight:

a) 3.0×10^7 N   **b)** 2.0×10^7 N   **c)** 8.0×10^6 N   **d)** 6.0×10^6 N   [2 marks each]

a) ....

b) ....

c) ....

d) ....

Hard questions: Scale diagrams. HT

Model example: An object has a 3 N force applied to it in a eastern direction and a 4 N force applied to it in a northern direction. Use a scale diagram to calculate the resultant force. [3 marks]

Set appropriate scale and draw eastern and northern forces: [1 mark]

Measure resultant force to be 5 N   [1 mark]
(range of 4.8–5.2 N allowed)

Give direction of force to be north-east   [1 mark]

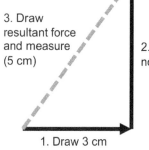

3. Draw resultant force and measure (5 cm)

2. Draw 4 cm northern force

1. Draw 3 cm eastern force

Q5. An object has a 8 N force exerted on it in a westerly direction, and a 3 N force exerted on it in a southerly direction. Using the space below, draw a scale diagram to calculate the resultant force.

[3 marks]

[Total marks   / 35]

Work done (combined science)

A force can cause an object to be moved a certain distance.

The equation that links work done, force and distance moved in the direction of the force is: $W = Fs$

You need to remember this equation!

Basic questions: No rearranging or unit conversion needed.

Model example: A constant force of 80 N is applied to a bowling ball over a distance of 0.5 m. Calculate the work done on the bowling ball. [3 marks]

Write down equation:	$W = Fs$	
Substitute variables into equation:	$W = 80 \times 0.5$	[1 mark]
Calculate answer:	$W = 80 \times 0.5 = 40$	[1 mark]
Write units:	joules (J)	[1 mark]

Q1. A constant force of 120 N is exerted on a pushchair. Calculate the work done on the pushchair for each of the distances:

a) 50 m   **b)** 10 m   **c)** 200 m   **d)** 40 m   [3 marks each]

a) ....

b) ....

c) ....

d) ....

Q2. A bicycle travels a distance of 800 m. Calculate the work done on the bicycle for each of the applied forces:

a) 20 N   **b)** 5 N   **c)** 300 N   **d)** 450 N   [3 marks each]

a) ....

b) ....

c) ....

d) ....

Medium questions: Rearranging and unit conversion needed (hints in boxes).

Q3. An object has a work done of 80 000 J applied to it. Calculate how far the object is moved for each of the forces.

kN → N × 1000

a) 2 kN   **b)** 6 kN   **c)** 0.5 kN   **d)** 0.8 kN   [4 marks each]

a) ....

b) ....

c) ....

d) ....

Q4. A crane applies a constant force of 2000 N in lifting an object. Calculate the distance that the object is lifted for each value of work done.

$kJ \rightarrow J \times 1000$

a) 20 kJ   **b)** 5 kJ   **c)** 50 kJ   **d)** 80 kJ   [4 marks each]

a) ....

b) ....

c) ....

d) ....

Hard questions: Rearranging, use of standard form and unit conversion needed.

Q5. An aeroplane does 3.6×10^{10} J of work in travelling a distance of 600 km. Calculate the force acting on the aeroplane. [4 marks]

....

....

Q6. A mountaineer has a weight of 800 N. Calculate the work done against gravity if the mountaineer climbs a mountain of height 6 km. [4 marks]

....

....

Mixed practice: Requires more than one equation.

Q7. Three people are pushing a broken down van. Two of the people exert a force of 800 N to the right, while the third exerts a force of 1.2 kN to the right.

a) Calculate the resultant force on the van. [3 marks]

....

....

b) The three people push the van a total distance of 1.4 km. Assuming that all of the forces are constant, calculate the work done in pushing the van. [4 marks]

....

....

[Total marks   / 71]

Forces and elasticity (combined science)

The extension of an elastic object, like a spring, is directly proportional to the force applied.
The equation that links force, spring constant and extension is: $F = ke$

> **You need to remember this equation!**

Basic questions: No rearranging or unit conversion needed.

Model example: A spring has a spring constant of 20 N/m and when a force is applied it is extended by a distance of 0.5 m. Calculate the force applied to the spring. [3 marks]

Write down equation:	$F = ke$	
Substitute variables into equation:	$F = 20 \times 0.5$	[1 mark]
Calculate answer:	$F = 20 \times 0.5 = 10$	[1 mark]
Write units:	newtons (N)	[1 mark]

Q1. An elastic band is extended by a distance of 0.2 m. Calculate the force applied to the elastic band for each spring constant:

a) 6 N/m   **b)** 2 N/m   **c)** 18 N/m   **d)** 8 N/m   [3 marks each]

a) ....

b) ....

c) ....

d) ....

Q2. A bungee cord has a spring constant of 120 N/m. Calculate the force applied to the bungee cord for each extension:

a) 5 m   **b)** 8 m   **c)** 20 m   **d)** 4 m   [3 marks each]

a) ....

b) ....

c) ....

d) ....

Medium questions: Rearranging and unit conversion needed (hints in boxes).

Q3. A spring has a spring constant of 20 000 N/m. Calculate the extension of the spring for each of the applied forces:

> kN → N × 1000

a) 2 kN   **b)** 10 kN   **c)** 2.5 kN   **d)** 8 kN   [4 marks each]

a) ....

b) ....

c) ....

d) ....

Q4. An elastic hair tie has a force of 1.8 N applied to it. Calculate the spring constant of the hair tie for each of the extensions:

$cm \rightarrow m \div 100$

a) 2 cm   **b)** 9 cm   **c)** 6 cm   **d)** 1.8 cm   [4 marks each]

a) ....

b) ....

c) ....

d) ....

Hard questions: Rearranging, use of standard form and unit conversion needed.

Q5. A spring has a spring constant 4×10^4 N/m. If the spring is compressed by a distance of 5 mm, calculate the force applied to the spring.   [4 marks]

....

....

Q6. A mattress contains springs of unknown spring constant. A force of 4 N is applied to one of the springs and it is compressed by a distance of 2 cm. Calculate the spring constant of the spring.   [4 marks]

....

....

Mixed practice: Requires more than one equation.

Q7. An elastic band of mass 2 g and spring constant 25 N/m is extended by 5 cm.

a) Calculate the force on the elastic band.   [4 marks]

....

....

b) Calculate the elastic potential energy stored by the elastic band.   [4 marks]

....

....

c) The elastic band is now released. Calculate the velocity of the elastic band immediately after it has been released.   [4 marks]

....

....

[Total marks   / 76]

Moments (physics only)

A moment is a turning force. The equation that links moment of a force, force and perpendicular distance to pivot is: $M = F\,d$

You need to remember this equation!

The principle of moments tells us a body is balanced when the sum of the clockwise moments about a point are equal to the sum of the anticlockwise moments about a point.

Basic questions: No rearranging or unit conversion needed.

Model example: A spanner of length 0.3 m is turning a bolt. A force of 5 N is applied to the end of the bolt. Calculate the moment. [3 marks]

Write down equation:	$M = F\,d$	
Substitute variables into equation:	$M = 5 \times 0.3$	[1 mark]
Calculate answer:	$M = 5 \times 0.3 = 1.5$	[1 mark]
Write units:	newton metres (Nm)	[1 mark]

Q1. A crowbar has a length of 0.8 m. Calculate the moment for each of the forces:

a) 3 N b) 8 N c) 10 N d) 7 N [3 marks each]

a) ....

b) ....

c) ....

d) ....

Medium questions: Determining whether or not a system is balanced.

Model example: Determine whether or not the system is balanced. [3 marks]

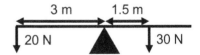

Calculate clockwise moments:	$M = 1.5 \times 30 = 45$ Nm	[1 mark]
Calculate anti-clockwise moments:	$M = 20 \times 3 = 60$ Nm	[1 mark]
State whether system is balanced:	Not balanced	[1 mark]

Q2. Determine whether or not each system is balanced:

a)
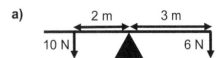

b)

c)
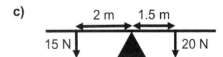

d)
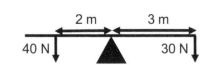

[3 marks each]

a) ....

....

b) ....

....

c) ....

....

d) ....

....

Hard questions: Determining whether or not a system is balanced (several forces).

Q3. Determine whether or not each system is balanced:

a)

| 2 m | 1.5 m | 1.5 m |

20 N | 10 N | 10 N

b)

| 1 m | 2 m | 1.5 m |

5 N | 15 N | 30 N

c)

| 3 m | 1.5 m | 1.5 m |

5 N | 2 N | 4 N

d)

| 2 m | 1.5 m | 1.5 m |

30 N | 20 N | 10 N

[5 marks each]

a) ....

....

....

b) ....

....

....

c) ....

....

....

d) ....

....

....

[Total marks / 44]

Pressure and pressure in a fluid (physics only)

Pressure is a measure of how much force is applied to every metre squared of area.

You need to remember this equation!

The equation that links pressure, force and area is: $p = \frac{F}{A}$

HT Objects submerged in a liquid also experience pressure as there is a column of liquid above the object.

This equation is given to you!

The equation that links pressure, height of the column, density of the liquid and gravitational field strength is $p = h\rho g$

Basic questions: Use of $p = \frac{F}{A}$

Model example: The bottom surface of a crate has an area of 0.5 m². The crate has a weight of 2000 N. Calculate the pressure the crate exerts on the floor. [3 marks]

Write down equation:	$p = \frac{F}{A}$	
Substitute variables into equation:	$p = \frac{2000}{0.5}$	[1 mark]
Calculate answer:	$p = \frac{2000}{0.5} = 4000$	[1 mark]
Write units:	pascals (Pa)	[1 mark]

Q1. The bottom surface of a shoe has a surface area of 0.02 m². Calculate the pressure exerted on the floor for each of the forces:

a) 100 N   **b)** 400 N   **c)** 250 N   **d)** 300 N   [3 marks each]

a) ....

b) ....

c) ....

d) ....

Medium questions: Use of $p = h\rho g$. Unit conversion needed (hint in box). **HT**

For all questions onwards in this chapter, take gravitational field strength to be 9.8 N/kg.

Model example: The density of a liquid is 1200 kg/m³. Calculate the pressure at a depth of 50 cm. [4 marks]

Convert unit:	h = 50 cm = 0.5 m	[1 mark]
Substitute variables into equation:	$p = h\rho g = 0.5 \times 1200 \times 9.8$	[1 mark]
Calculate answer:	$p = h\rho g = 0.5 \times 1200 \times 9.8 = 5900$	[1 mark]
Write units:	pascals (Pa)	[1 mark]

Q2. Saltwater has a density of 1020 kg/m^3. Calculate the pressure on an object at the following depths:

cm $\rightarrow$ m $\div$ 100

a) 20 cm   **b)** 50 cm   **c)** 120 cm   **d)** 85 cm   [4 marks each]

a) ....

b) ....

c) ....

d) ....

Hard questions: Use of both equations. Rearranging and unit conversion needed. HT

Q3. Calculate the force needed to exert a pressure of 50 000 Pa onto each area:

a) 5 cm^2 **b)** 20 cm^2 **c)** 1000 mm^2 **d)** 2500 mm^2 [4 marks each]

a) ....

b) ....

c) ....

d) ....

Q4. An object is experiencing pressure while submerged in a liquid of density 1050 kg/m^3. Calculate how far the object is under the surface of the liquid for each pressure:

a) 50 kPa   **b)** 40 kPa   **c)** 15 kPa   **d)** 22 kPa   [4 marks each]

a) ....

b) ....

c) ....

d) ....

Mixed practice: Requires more than one equation.

Q5. A liquid has a mass of 3000 kg and a volume of 6 m^3. Calculate the pressure exerted on an object that is submerged at a depth of 40 cm within the liquid.   [6 marks]

....

....

....

[Total marks   / 66]

Speed (combined science)

Speed is a measure of how far an object travels in a certain time.

The equation that links distance travelled, speed and time is: $s = v\,t$

You need to remember this equation!

Basic questions: No rearranging or unit conversion needed.

Model example: A student is walking at a speed of 1.5 m/s. If they walk at this speed for a time of 30 seconds, calculate how far they walk. [3 marks]

Write down equation:	$s = v\,t$	
Substitute variables into equation:	$s = 1.5 \times 30$	[1 mark]
Calculate answer:	$s = 1.5 \times 30 = 45$	[1 mark]
Write units:	metres (m)	[1 mark]

Q1. Calculate how far somebody jogs if they jog at a speed of 3 m/s for the following times:

a) 20 s   b) 50 s   c) 120 s   d) 150 s   [3 marks each]

a) ....

b) ....

c) ....

d) ....

Q2. A cyclist is travelling at a speed of 6 m/s. Calculate how far they travel if they cycle for the following times:

a) 20 s   b) 50 s   c) 120 s   d) 150 s   [3 marks each]

a) ....

b) ....

c) ....

d) ....

Medium questions: Rearranging and unit conversion needed (hints in boxes).

Q3. A car travels a distance of 1.2 km. Calculate the speed of the car if it travels this distance in the following times:

km → m × 1000

a) 45 s   b) 30 s   c) 50 s   d) 90 s   [4 marks each]

a) ....

b) ....

c) ....

d) ....

Q4. A motorbike travels a distance of 3000 m. Calculate the speed of the motorbike if it travels this distance in the following times:

mins → s ÷ 60

a) 3 minutes   **b)** 2 minutes   **c)** 8 minutes   **d)** 5 minutes   [4 marks each]

a) ....

b) ....

c) ....

d) ....

Hard questions: Rearranging and unit conversion needed.

Q5. A sports car travels a distance of 110 km in a time of one hour. Calculate the speed of the car in metres per second. [4 marks]

....

....

Q6. An aeroplane travels a distance of 300 km in a time of 30 minutes. Calculate the speed of the aeroplane in metres per second. [4 marks]

....

....

Q7. A ferry travels a distance of 40 km in a time of two hours. Calculate the speed of the ferry in metres per second. [4 marks]

....

....

Mixed practice: Requires more than one equation.

Q8. A car has a mass of 1200 kg and travels a distance of 20 km in a time of 15 minutes. Calculate the average kinetic energy of the car. [6 marks]

....

....

....

[Total marks   / 74]

Distance–time graphs (combined science)

> The gradient of a distance–time graph gives us the speed of an object. The steeper the gradient, the faster the speed.
>
> **You need to remember this!**
>
> If the gradient of a distance–time graph is zero, then the object is stationary. If the distance–time graph is a straight line (constant gradient) then the object is travelling at a constant speed. If the gradient is increasing then the object is accelerating and if the gradient is decreasing then the object is decelerating.

Basic questions: Obtaining distances travelled and speeds from distance–time graphs

Model example: A distance–time graph for a cyclist is shown.
Calculate the speed of the cyclist.   [3 marks]

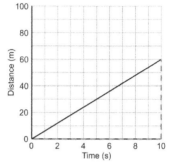

Draw gradient triangle.   [1 mark]

Calculate gradient: $\frac{60}{10} = 6$ [1 mark]

Write units: metres per second (m/s)   [1 mark]

Q1. Using the distance–time graphs, state how far each object has travelled:

a)

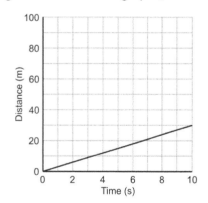

b)   [1 mark each]

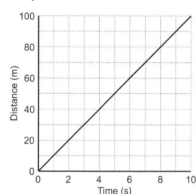

a) ....

b) ....

Q2. Using the same distance–time graphs as question 1, calculate the speed of each object. [3 marks each]

a) ....

....

b) ....

....

Medium questions: Rearranging and unit conversion needed (hint in box).

Q3. Use the distance–time graphs of a fighter jet to calculate its speed in metres/second.

$$km \rightarrow m \times 1000$$

a)

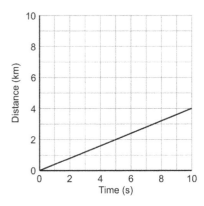

b)

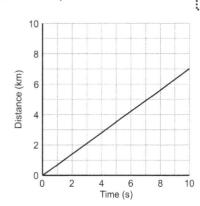

[3 marks each]

a) ....

....

b) ....

....

Hard questions: Obtaining speed from tangents of a curve. **HT**

Q4. Using the graph on the right, find the speed
(in units of metres/second) of the object at each time:

a) 8 minutes

b) 15 minutes

c) 20 minutes   [4 marks each]

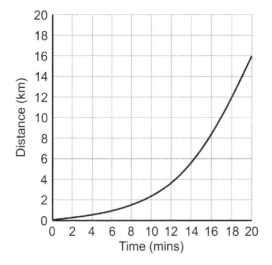

a) ....

....

b) ....

....

c) ....

....

[Total marks   / 26]

Acceleration (combined science)

Acceleration is a measure of how much the velocity changes in a given time.

The equation that links acceleration, change in velocity and time is: $a = \frac{\Delta v}{t}$

> **You need to remember this equation!**

Basic questions: No rearranging or unit conversion needed.

Model example: An athlete increases their velocity by 4 m/s in a time of 2 s. Calculate the acceleration of the athlete. **[3 marks]**

Write down equation: $\qquad a = \frac{\Delta v}{t}$

Substitute variables into equation: $\qquad a = \frac{4}{2}$ **[1 mark]**

Calculate answer: $\qquad a = \frac{4}{2} = 2$ **[1 mark]**

Write units: $\qquad$ metres per second squared (m/s^2) **[1 mark]**

Q1. A car increases its velocity by 20 m/s. Calculate the acceleration of the car if this velocity increase happens in the following times:

a) 10 s $\qquad$ **b)** 4 s $\qquad$ **c)** 30 s $\qquad$ **d)** 12 s $\qquad$ [3 marks each]

a) ....

b) ....

c) ....

d) ....

Medium questions: Calculating changes in velocity and unit conversion needed (hints in boxes).

Q2. A scooter accelerates from a velocity of 5 m/s to a velocity of 9 m/s in a time of 8 seconds.

a) Calculate the scooter's change in velocity. **[1 mark]**

....

b) Calculate the acceleration of the scooter. **[3 marks]**

....

....

Q3. An oil tanker accelerates from a velocity of 0.5 m/s to a velocity of 2 m/s. Calculate the acceleration of the oil tanker for each of the times:

> minutes → s × 60

a) 3 minutes $\qquad$ **b)** 6 minutes $\qquad$ **c)** 5 minutes $\qquad$ **d)** 1 minute $\qquad$ [4 marks each]

a) ....

b) ....

c) ....

d) ....

Q4. A electron is initially at rest and is accelerated for a time of 2 seconds. Calculate the acceleration for each final velocity:

km/s → m/s × 1000

a) 80 km/s   b) 120 km/s   c) 20 km/s   d) 300 km/s   [4 marks each]

a) ....

b) ....

c) ....

d) ....

Hard questions: Rearranging, use of standard form and unit conversion needed.

Q5. A proton is initially at rest and then experiences an acceleration of 4×10^5 m/s^2. Calculate the final velocity of the proton for each of the acceleration times:

a) 200 ms   b) 40 ms   c) 120 ms   d) 800 ms   [4 marks each]

a) ....

b) ....

c) ....

d) ....

Q6. A fighter jet accelerates from rest to a velocity of 1.2 km/s. For each acceleration, calculate the time taken to get to this velocity.

a) 20 m/s^2 b) 25 m/s^2 c) 30 m/s^2 d) 5 m/s^2 [4 marks each]

a) ....

b) ....

c) ....

d) ....

Mixed practice: Requires more than one equation.

Q7. A footballer of mass 65 kg is initially running at a velocity of 3 m/s and then accelerates at 2 m/s^2 for a time of 2 seconds. Calculate the kinetic energy of the footballer after they have stopped accelerating.   [5 marks]

....

....

[Total marks   / 85]

Velocity–time graphs (combined science)

The gradient of a velocity–time graph gives us the acceleration of an object. The steeper the gradient, the greater the acceleration. Note that a negative acceleration is commonly referred to as a deceleration.
The area underneath a velocity–time graph gives the total distance travelled.

> **You need to remember this!**

Basic questions: Calculating acceleration from velocity–time graphs.

Model example: A velocity–time graph for a rugby player is shown. Calculate the acceleration of the rugby player. [3 marks]

Draw gradient triangle. [1 mark]

Calculate gradient: $-8 \div 4 = -2$ [1 mark]

Write units: metres per second squared (m/s^2) [1 mark]

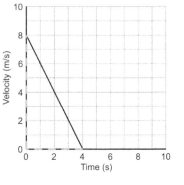

Q1. Using the velocity–time graphs, calculate the acceleration of each object:

a) 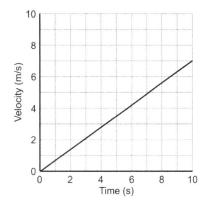 **b)** 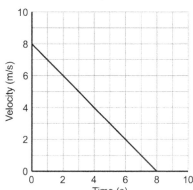 [3 marks each]

a) ....

....

....

b) ....

....

....

Medium questions: Calculating distance travelled from velocity–time graphs. HT

Model example: A velocity–time graph for a bicycle is shown. Calculate the distance travelled by the bicycle. [3 marks]

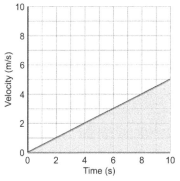

Work out the area under the line (here, a triangle). $\frac{1}{2} \times 10 \times 5$ [1 mark]

Calculate area: $\frac{1}{2} \times 10 \times 5 = 25$ [1 mark]

Write units: metres (m) [1 mark]

Q2. Using the graphs in question 1, calculate the distance travelled by each object: [3 marks each]

a) ....

....

b) ....

....

Hard questions: Calculating distance travelled from more complex velocity–time graphs. HT

Q3. Using the velocity–time graphs, calculate the distance travelled by each object:

a) 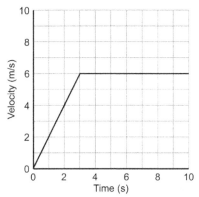   b) 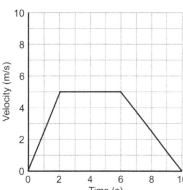 [3 marks each]

a) ....

....

b) ....

....

Q4. Using the graph in question 3a), calculate the acceleration of the object between 0 and 3 seconds. [3 marks]

....

....

[Total marks   / 21]

Equation of motion (combined science)

This equation is for cases where there is uniform (constant) acceleration and links final velocity, initial velocity, acceleration and distance travelled: $v^2 - u^2 = 2\,a\,s$

> This equation is given to you!

Basic questions: Initial velocity of zero, no rearranging needed.

Model example: An athlete starts a race from rest. They accelerate at 1.2 m/s^2 over a distance of 40 m. Calculate the velocity of the athlete at 40 m. [3 marks]

Write down equation:	$v^2 - u^2 = 2\,a\,s$	
Substitute variables into equation:	$v^2 - 0^2 = 2 \times 1.2 \times 40$	[1 mark]
Calculate answer:	$v^2 = 96,\ v = 9.8$	[1 mark]
Write units:	metres per second (m/s)	[1 mark]

Q1. A hockey player starts from rest and accelerates at 0.8 m/s^2. Calculate the velocity of the hockey player after each distance:

a) 10 m   **b)** 25 m   **c)** 40 m   **d)** 22 m   [3 marks each]

a) ....

....

b) ....

....

c) ....

....

d) ....

....

Medium questions: Rearranging needed.

Model example: A cheetah is running at 10 m/s and accelerates at 4 m/s^2 over a distance of 60 m. Calculate the velocity of the cheetah at 60 m. [3 marks]

Substitute variables into equation:	$v^2 - 10^2 = 2 \times 4 \times 60$	[1 mark]
Calculate answer:	$v^2 = 10^2 + 2 \times 4 \times 60,\ v^2 = 580,\ v = 24$	[1 mark]
Write units:	metres per second (m/s)	[1 mark]

Q2. A helicopter is travelling at 20 m/s and accelerates at 5 m/s^2. Calculate the velocity of the helicopter after each distance:

a) 80 m   **b)** 300 m   **c)** 120 m   **d)** 500 m   [3 marks each]

a) ....

....

b) ....

....

c) ....

....

d) ....

Hard questions: Rearranging and unit conversion needed.

Q3. An oil tanker accelerates from 0.5 m/s to 2.5 m/s. Calculate the acceleration of the oil tanker for each distance travelled:

a) 2 km   **b)** 3 km   **c)** 5 km   **d)** 0.8 km   [4 marks each]

a) ....

....

b) ....

....

c) ....

....

d) ....

....

Mixed practice: Requires more than one equation.

Q4. A rugby player accelerates from rest and reaches a top speed of 9 m/s in a time of 3 seconds. Calculate the distance that the rugby player travelled in this time.   [5 marks]

....

....

....

[Total marks   / 45]

Newton's second law (combined science)

Newton's second law says that the resultant force applied to an object is directly proportional to its acceleration. It also says that the acceleration of an object is inversely proportional to its mass.

The equation for Newton's second law links force, mass and acceleration: $F = m\,a$

You need to remember this equation!

Basic questions: No rearranging or unit conversion needed.

Model example: A car has a mass of 1200 kg and accelerates at 0.6 m/s². Calculate the resultant force acting on the car. [3 marks]

Write down equation:	$F = m\,a$	
Substitute variables into equation:	$F = 1200 \times 0.6$	[1 mark]
Calculate answer:	$F = 1200 \times 0.6 = 720$	[1 mark]
Write units:	newtons (N)	[1 mark]

Q1. A van has a mass of 2500 kg. Calculate the resultant force acting on the van for each of the accelerations:

a) 2 m/s²   **b)** 1.6 m/s²   **c)** 0.5 m/s²   **d)** 2.4 m/s²   [3 marks each]

a) ....

b) ....

c) ....

d) ....

Medium questions: Rearranging needed.

Q2. A bicycle and its rider have a combined mass of 80 kg. Calculate the acceleration for each of the resultant forces:

a) 160 N   **b)** 240 N   **c)** 20 N   **d)** 640 N   [3 marks each]

a) ....

b) ....

c) ....

d) ....

Q3. A remote controlled car has a resultant force of 4 N acting on it. Calculate the mass for each acceleration:

a) 6 m/s²   **b)** 8 m/s²   **c)** 5 m/s²   **d)** 2 m/s²   [3 marks each]

a) ....

b) ....

c) ....

d) ....

Hard questions: Rearranging and unit conversion needed.

Q4. A wooden block has a resultant force of 5 N applied to it. Calculate the acceleration for each mass:

    a) 800 g   **b)** 500 g   **c)** 1200 g   **d)** 700 g   [4 marks each]

a) ....

b) ....

c) ....

d) ....

Q5. A car has a mass of 1500 kg. Calculate the acceleration of the car for each resultant force:

    a) 2 kN   **b)** 1.4 kN   **c)** 3 kN   **d)** 0.8 kN   [4 marks each]

a) ....

b) ....

c) ....

d) ....

Mixed practice: Requires more than one equation.

Q6. A sprinter of mass 80 kg accelerates from rest to a velocity of 10 m/s in a time of 4 seconds. Calculate the resultant force that causes this acceleration.   [5 marks]

....

....

....

[Total marks   / 73]

Stopping distance (combined science)

The overall distance that a vehicle takes to stop is called the stopping distance. This stopping distance is made up of two parts: the thinking distance and the braking distance. The thinking distance is how far the vehicle travels before the driver is able to react (during the reaction time). The braking distance is how far the vehicle travels while the brakes are applied.

> **You need to remember this equation!**

Stopping distance = *thinking distance* + *braking distance*

Basic questions: Calculation of stopping distances.

Q1. A car brakes to avoid a collision. The thinking distance is 9 m. Calculate the stopping distance for each braking distance:

a) 14 m   **b)** 20 m   [1 mark each]

a) ....

b) ....

Q2. A motorbike brakes to avoid a collision. The thinking distance is 15 m. Calculate the braking distance for each stopping distance:

a) 38 m   **b)** 45 m   [1 mark each]

a) ....

b) ....

Q3. A motorbike brakes to avoid a collision. The braking distance is 55 m. Calculate the thinking distance for each stopping distance:

a) 73 m   **b)** 80 m   [1 mark each]

a) ....

b) ....

Medium questions: Use of $s = vt$ needed. Calculating thinking distance from reaction time.

Q4. A driver is travelling at a velocity of 20 m/s. Calculate the thinking distance for each reaction time:

> $ms \rightarrow s \div 1000$

a) 120 ms   **b)** 180 ms   **c)** 200 ms   **d)** 250 ms   [4 marks each]

a) ....

b) ....

c) ....

d) ....

Q5. Looking at your answers to question 4. State what happens to the thinking distance as the reaction time increases. [1 mark]

....

Hard questions: Use of $v^2 - u^2 = 2\,a\,s$ needed. Calculating braking distance from deceleration.

Q6. A van is initially travelling at 15 m/s and, while the brakes are pressed, decelerates to rest. Calculate the braking distance of the van for each deceleration:

a) 1.2 m/s^2 **b)** 1.4 m/s^2 **c)** 0.8 m/s^2 **d)** 0.5 m/s^2 [4 marks each]

a) ....

b) ....

c) ....

d) ....

Mixed practice: Requires more than one equation.

Q7. These questions involve a car of mass 1100 kg.

a) The car is initially travelling at 10 m/s. Calculate its kinetic energy. [3 marks]

....

....

b) The car then accelerates to 20 m/s. Calculate its kinetic energy. [3 marks]

....

....

c) Using your answers to parts a) and b). State what happens to the kinetic energy as the velocity of a vehicle is doubled. [1 mark]

....

d) State the equation that links work done, force and distance. [1 mark]

....

e) The car brakes under a constant force. Using your answers to parts c) and d) state and explain what happens to the braking distance of a car if the velocity is doubled. [2 marks]

....

....

[Total marks   / 49]

Momentum (combined science) HT

Momentum is equal to the mass of an object multiplied by its velocity.

You need to remember this equation!

Conservation of momentum says that the total momentum before an event (for example a collision) must be equal to the total momentum after an event.

The equation for momentum is: $p = m\,v$

Basic questions: No rearranging or unit conversion needed.

Model example: A football has a mass of 0.4 kg and is travelling at a velocity of 12 m/s. Calculate the momentum of the football. [3 marks]

Write down equation:	$p = m\,v$	
Substitute variables into equation:	$p = 0.4 \times 12$	[1 mark]
Calculate answer:	$p = 0.4 \times 12 = 4.8$	[1 mark]
Write units:	kilograms metre per second (kg m/s)	[1 mark]

Q1. A bowling ball has a mass of 5 kg. Calculate the momentum of the bowling ball for each of the velocities:

a) 2 m/s   b) 3.2 m/s   c) 2.4 m/s   d) 1.8 m/s   [3 marks each]

a) ....

b) ....

c) ....

d) ....

Medium questions: Rearranging and unit conversion needed (hints in boxes).

Q2. A tennis ball has a mass of 56 g. Calculate the velocity of tennis ball for each of the momentums:

g → kg ÷ 1000

a) 2.2 kg m/s   b) 2.4 kg m/s   c) 1.6 kg m/s   d) 0.9 kg m/s   [4 marks each]

a) ....

b) ....

c) ....

d) ....

Q3. Different fighter jets each have a momentum of 3×10^7 kg m/s. Calculate the mass of each fighter jet for each of the velocities:

km/s → m/s × 1000

a) 1 km/s   b) 1.2 km/s   c) 0.7 km/s   d) 0.8 km/s   [4 marks each]

a) ....

b) ....

c) ....

d) ....

Hard questions: Conservation of momentum questions.

Model example: A paintball of mass 3 g is fired with a velocity of 80 m/s from a barrel of mass 2 kg. Calculate the recoil velocity of the paintball machine.   [5 marks]

Convert unit:	$3\ g = 0.003\ kg$	[1 mark]
Calculate momentum of paintball:	$p = m\,v = 0.003 \times 80 = 0.24$	[1 mark]
This momentum is equal and opposite to momentum of barrel. Substitute variables into equation:	$p = m\,v$ $0.24 = 2 \times v$	[1 mark]
Calculate answer:	$v = 0.24 \div 2 = 0.12$	[1 mark]
Write units:	metre per second (m/s)	[1 mark]

Q4. A paintball of mass 1.2 g is fired from a barrel of mass 1.5 kg. Calculate the recoil velocity of the barrel for each paintball velocity:

a) 75 m/s   **b)** 90 m/s   **c)** 105 m/s   **d)** 110 m/s   [5 marks each]

a) ....

....

b) ....

....

c) ....

....

d) ....

....

Q5. A car of mass 1200 kg is travelling at a velocity of 15 m/s when it collides with a stationary van of mass 2000 kg. After the collision, the car and the van stick together. Calculate the velocity of the car and van immediately after the collision.   [5 marks]

....

....

[Total marks   / 69]

Changes in momentum (physics only) HT

The equation $a = \frac{\Delta v}{t}$ can be substituted into $F = m\,a$ to give a new equation that relates force, mass, change in velocity and time: $F = \frac{m\,\Delta v}{\Delta t}$.

This means that force is equal to the rate of change of momentum.

This equation is given to you!

Basic questions: No rearranging or unit conversion needed.

Model example: A car of mass 1400 kg crashes and experiences a change in velocity of 5 m/s in a time of 0.5 seconds. Calculate the force experienced by the car. [3 marks]

Write down equation: $\qquad\qquad\qquad F = \frac{m\,\Delta v}{\Delta t}$

Substitute variables into equation: $\qquad F = \frac{1400 \times 5}{0.5}$ [1 mark]

Calculate answer: $\qquad\qquad\qquad F = \frac{1400 \times 5}{0.5} = 14\,000$ [1 mark]

Write units: $\qquad\qquad\qquad\qquad\qquad$ newtons (N) [1 mark]

Q1. A van of mass 2200 kg crashes and experiences a change in velocity of 12 m/s. Calculate the force experienced by the van for each collision time:

a) 0.05 s $\qquad$ b) 0.3 s $\qquad$ c) 0.4 s $\qquad$ d) 0.7 s $\qquad$ [3 marks each]

a) ....

b) ....

c) ....

d) ....

Q2. a) State what happens to the force when the collision time increases. [1 mark]

....

b) Explain, in terms of rate of change of momentum, how air bags reduce the chance of injury. [2 marks]

....

....

Medium questions: Rearranging and unit conversion needed (hints in boxes).

Q3. A cricket ball of mass 160 g experiences a change in velocity of 40 m/s. Calculate the contact time for each force:

$g \rightarrow kg \div 1000$

a) 640 N $\qquad$ b) 530 N $\qquad$ c) 800 N $\qquad$ d) 256 N $\qquad$ [4 marks each]

a) ....

b) ....

c) ....

d) ....

Q4. A tennis ball of mass 56 g experiences a force of 400 N. Calculate the change in velocity for each contact time:

$ms \rightarrow s \div 1000$

a) 5 ms   **b)** 4 ms   **c)** 6 ms   **d)** 2.4 ms   [5 marks each]

a) ....

b) ....

c) ....

d) ....

Hard questions: Rearranging and unit conversion needed.

Q5. A rugby player has a mass of 90 kg and is initially travelling at 8 m/s. The rugby player is tackled with a force of 1.6 kN and comes to a complete stop. Calculate the contact time of the tackle.   [4 marks]

....

....

Q6. A golf ball of mass 46 g is hit by a club and experiences a force of 2.5 kN which causes a change in velocity of 50 m/s. Calculate the contact time between club and ball.   [5 marks]

....

....

Mixed practice: Requires more than one equation.

Q7. A brick has a density of 2000 kg/m^3 and a volume of 1500 cm^3. The brick is dropped and hits the ground. It experiences a change in velocity of 8 m/s in a time of 2 ms. Calculate the force experienced by the brick.   [6 marks]

....

....

....

....

[Total marks   / 66]

Frequency and time period (combined science)

Frequency is a measure of how many waves pass a point every second. The time period is the overall time for a complete wave to pass a point.

The equation that links time period and frequency is: $T = \frac{1}{f}$

This equation is given to you!

Basic questions: No rearranging or unit conversion needed.

Model example: A wave has a frequency of 5 Hz. Calculate the time period of the wave. [3 marks]

Write down equation: $T = \frac{1}{f}$

Substitute variables into equation: $T = \frac{1}{5}$ [1 mark]

Calculate answer: $T = \frac{1}{5} = 0.2$ [1 mark]

Write units: seconds (s) [1 mark]

Q1. Calculate the time period of a wave for each frequency:

a) 0.2 Hz **b)** 50 Hz **c)** 100 Hz **d)** 4 Hz [3 marks each]

a) ....

b) ....

c) ....

d) ....

Medium questions: Rearranging and unit conversion needed (hints in boxes).

Q2. Calculate the time period of a wave for each frequency:

kHz → Hz × 1000

a) 4 kHz **b)** 20 kHz **c)** 0.5 kHz **d)** 250 kHz [4 marks each]

a) ....

b) ....

c) ....

d) ....

Q3. Calculate the time period of a wave for each frequency:

MHz → Hz × 1 000 000

a) 1 MHz **b)** 12 MHz **c)** 5 MHz **d)** 30 MHz [4 marks each]

a) ....

b) ....

c) ....

d) ....

Q4. Calculate the frequency of a wave for each time period: $\boxed{\text{ms} \rightarrow \text{s} \div 1000}$

    a) 2 ms   **b)** 40 ms   **c)** 5 ms   **d)** 250 ms   [4 marks each]

    a) ....

    b) ....

    c) ....

    d) ....

Q5. Calculate the frequency of a wave for each time period: $\boxed{\mu\text{s} \rightarrow \text{s} \div 1\,000\,000}$

    a) 200 μs   **b)** 120 μs   **c)** 50 μs   **d)** 18 μs   [4 marks each]

    a) ....

    b) ....

    c) ....

    d) ....

Hard questions: Rearranging and unit conversion needed.

Q6. A radio station broadcasts at a frequency of 98 MHz. Calculate the time period of these radio waves.   [4 marks]

....

....

Q7. Microwaves are emitted at a frequency of 300 GHz. Calculate the time period of the microwaves. $\boxed{\text{GHz} \rightarrow \text{Hz} \times 10^9}$

[4 marks]

....

....

Q8. A wave has a time period of 4 ms. Calculate how many waves pass a point in one second.   [3 marks]

....

....

[Total marks / 87]

Wave speed equation (combined science)

Wave speed is a measure of how far a wave travels each second.

The equation that links wave speed, frequency and wavelength is: $v = f\lambda$

You need to remember this equation!

Basic questions: No rearranging or unit conversion needed.

Model example: A sound wave has a frequency of 680 Hz and a wavelength of 0.5 m. Calculate the speed of the sound wave. [3 marks]

Write down equation: $v = f\lambda$

Substitute variables into equation: $v = 680 \times 0.5$ [1 mark]

Calculate answer: $v = 680 \times 0.5 = 340$ [1 mark]

Write units: metres per second (m/s) [1 mark]

Q1. A wave has a frequency of 200 Hz. Calculate the speed of the wave for each of the wavelengths:

    a) 0.4 m   **b)** 12 m   **c)** 1.5 m   **d)** 20 m   [3 marks each]

    a) ....

    b) ....

    c) ....

    d) ....

Medium questions: Rearranging and unit conversion needed (hints in boxes).

Q2. A sound wave has a speed of 340 m/s. Calculate the wavelength of the sound wave for each frequency:

    kHz → Hz × 1000

    a) 2 kHz   **b)** 0.6 kHz   **c)** 5 kHz   **d)** 12 kHz   [4 marks each]

    a) ....

    b) ....

    c) ....

    d) ....

Q3. The speed of sound is higher in solids than in air. A sound wave has a wavelength of 0.4 m in a solid. Calculate the frequency for each speed:

    km/s → m/s × 1000

    a) 2 km/s   **b)** 2.8 km/s   **c)** 1.2 km/s   **d)** 0.8 km/s   [4 marks each]

a) ....

b) ....

c) ....

d) ....

Hard questions: Rearranging, use of standard form and unit conversion needed.

Q4. Electromagnetic waves travel at 3.0×10^8 m/s. Calculate the frequency for each wavelength:

    a) A radio wave of wavelength 20 cm   **b)** A microwave of wavelength 0.8 mm

    c) A UV wave of wavelength 0.05 μm   **d)** Green light of wavelength 532 nm   [4 marks each]

a) ....

b) ....

c) ....

d) ....

Q5. A radio wave has a frequency of 15 MHz. Calculate the wavelength of the radio wave.   [4 marks]

....

....

Q6. A microwave has a frequency of 900 GHz. Calculate the wavelength of the microwave.   [4 marks]

....

....

Mixed practice: Requires more than one equation.

Q7. An electromagnetic wave travelling at 3.0×10^8 m/s has a time period of 3.5×10^{-15} s.
Calculate the wavelength of the electromagnetic wave.   [5 marks]

....

....

Q8. A wave travels a distance of 2 km in 40 s. The frequency of the wave is 200 Hz. Calculate the wavelength.   [6 marks]

....

....

[Total marks   / 79]

Magnification (physics only)

Light incident onto a lens is refracted and forms an image. If the image is larger than the size of an object it is said to be magnified. The equation that links image height and object height is: $magnification = \frac{image\ height}{object\ height}$

This equation is given to you!

Basic questions: No rearranging or unit conversion needed.

Model example: The object height is 0.002 m and the image height is 0.05 m, calculate the magnification. [2 marks]

Write down equation: $magnification = \frac{image\ height}{object\ height}$

Substitute variables into equation: $magnification = \frac{0.05}{0.002}$ [1 mark]

Calculate answer: $magnification = \frac{0.05}{0.002} = 25$ [1 mark]

Q1. A convex lens is imaging an object of height 0.003 m. Calculate the magnification for each image height:

    a) 0.06 m   **b)** 0.15 m   **c)** 0.045 m   **d)** 0.09 m   [2 marks each]

    a) ....

    b) ....

    c) ....

    d) ....

Q2. A magnifying glass creates an image of height 0.04 m. Calculate the magnification for each object height:

    a) 0.002 m   **b)** 0.005 m   **c)** 0.008 m   **d)** 0.012 m   [2 marks each]

    a) ....

    b) ....

    c) ....

    d) ....

Medium questions: Rearranging and unit conversion needed (hints in boxes).

Q3. A microscope creates an image of a cell of height 0.002 m. Calculate the magnification for each object height:

$\mu m \rightarrow m \div 1\ 000\ 000$

    a) 40 μm   **b)** 50 μm   **c)** 70 μm   **d)** 10 μm   [3 marks each]

    a) ....

    b) ....

c) ....

d) ....

Q4. A magnifying glass has a magnification of 6. Calculate the image height (in m) for each object height:

$mm \rightarrow m \div 1000$

    a) 3 mm   **b)** 5 mm   **c)** 8 mm   **d)** 14 mm   [3 marks each]

a) ....

b) ....

c) ....

d) ....

Q5. A projector produces an image height of 200 cm. Calculate the object height (in m) for each magnification:

$cm \rightarrow m \div 100$

    a) 50   **b)** 12   **c)** 15   **d)** 40   [3 marks each]

a) ....

b) ....

c) ....

d) ....

Hard questions: Rearranging and unit conversion needed.

Q6. A convex lens is imaging an object of height 5 mm. The image height is 20 cm. Calculate the magnification. [3 marks]

....

....

Q7. A chloroplast has a length of 5 µm. Calculate the magnification if an image of a chloroplast has a length of 1 cm. [3 marks]

....

....

Q8. A cell membrane has a thickness of 7 nm. In a diagram it is shown with a thickness of 1 mm. Calculate the magnification of the diagram. [3 marks]

....

....

[Total marks   / 61]

Force on a current carrying wire (combined science) HT

A wire carrying a current produces a magnetic field. A current carrying wire at right angles to a magnetic field experiences a force.

The equation that links force, magnetic flux density, current and length is: $F = BIl$

This equation is given to you!

Basic questions: No rearranging or unit conversion needed.

Model example: A current of 2 A is flowing through a wire. A wire of length 0.1 m is at right angles to a magnetic field of flux density 0.04 T. Calculate the force acting on the wire.　　[3 marks]

Write down equation:

$$F = BIl$$

Substitute variables into equation:

$$F = 0.04 \times 2 \times 0.1$$　　[1 mark]

Calculate answer:

$$F = 0.04 \times 2 \times 0.1 = 0.008$$　　[1 mark]

Write units:

newtons (N)　　[1 mark]

Q1. A current of 0.5 A is flowing through a wire. The wire is at right angles to a magnetic field 0.1 m in length. Calculate the force for each magnetic flux density:

a) 0.2 T　　　　**b)** 0.05 T　　　　**c)** 0.09 T　　　　**d)** 0.12 T　　　　[3 marks each]

a) ....

b) ....

c) ....

d) ....

Q2. A wire is travelling at right angles through a magnetic field of flux density 0.04 T and length 0.15 m. Calculate the force on the wire for each current:

a) 5 A　　　　**b)** 2 A　　　　**c)** 0.4 A　　　　**d)** 13 A　　　　[3 marks each]

a) ....

b) ....

c) ....

d) ....

Medium questions: Rearranging and unit conversion needed (hints in boxes).

Q3. A force of 0.08 N is acting on a current carrying wire at right angles to a magnetic field of flux density 0.05 T. Calculate the current for each length of wire:

cm → m ÷ 100

a) 15 cm　　　　**b)** 20 cm　　　　**c)** 50 cm　　　　**d)** 4 cm　　　　[4 marks each]

a) ....

b) ....

c) ....

d) ....

Q4. A wire has a current of 2 A flowing through it at right angles to a magnetic field of length 0.5 m. Calculate the magnetic flux density for each force acting on the wire:

$mN \rightarrow N \div 1000$

a) 1.2 mN	**b)** 20 mN	**c)** 400 mN	**d)** 15 mN	[4 marks each]

a) ....

b) ....

c) ....

d) ....

Hard questions: Rearranging and unit conversion needed.

Q5. A wire carrying a current of 1.2 A is at right angles to a magnetic field of flux density 4 mT. If there is a force of 0.02 N acting on the wire, calculate the length of the wire. [4 marks]

....

....

Q6. A wire of length 20 cm is carrying a current of 0.4 A. The wire is at right angles to a magnetic field and experiences a force of 0.6 N. Calculate the magnetic flux density of the magnetic field. [4 marks]

....

....

Mixed practice: Requires more than one equation.

Q7. A wire of length 0.2 m carries a current of 4 A at right angles to a magnetic field of flux density 8×10^{-2} T. The wire floats in the magnetic field. Calculate the mass of the wire. Take the gravitational field strength to be 9.8 N/kg. [5 marks]

....

....

Q8. A conductor has resistance of 2.4 kΩ and a potential difference of 12 V across it. The conductor has a length of 4 cm and experiences a force of 0.01 N when at right angles to a magnetic field. Calculate the magnetic flux density of the magnetic field. [6 marks]

....

....

[Total marks   / 75]

Transformers (physics only) HT

A step-up transformer increases the potential difference of a supply, and decreases the current. A step-down transformer does the opposite.

These equations are given to you!

Assuming 100% efficiency, the equation that links power output from the secondary coil of a transformer and the power input to the primary coil is: $V_p \times I_p = V_s \times I_s$ (Combined science covers this equation only)

The equation that links potential difference across the primary and secondary coils and the number of turns in the primary and secondary coils is: $\frac{V_p}{V_s} = \frac{n_p}{n_s}$

Basic questions: Use of $V_p \times I_p = V_s \times I_s$

Model example: There is a current of I_s = 2 A in and potential difference of V_s = 240 V across the secondary coil of a transformer. The primary coil has a potential difference of V_p = 12 V across it. Calculate the current in the primary coil.   [3 marks]

Write down equation: $V_p \times I_p = V_s \times I_s$

Substitute variables into equation: $12 \times I_p = 240 \times 2$ [1 mark]

Rearrange equation and calculate answer: $I_p = \frac{240 \times 2}{12} = \frac{480}{12} = 40$ [1 mark]

Write units: amps (A)   [1 mark]

Q1. There is a current of I_s = 0.5 A in and potential difference of V_s = 120 V across the secondary coil of a transformer. Calculate the current in the primary coil for each potential difference across the primary coil:

    a) V_p = 6 V   **b)** V_p = 1.5 V   **c)** V_p = 1 V   **d)** V_p = 40 V   [3 marks each]

    a) ....

    b) ....

    c) ....

    d) ....

Q2. There is a current of I_s = 2 A in the secondary coil of a transformer, and a current I_p = 0.1 A in the primary coil of a transformer. Calculate the potential difference across the secondary coil for each primary coil potential difference:

    a) V_p = 3 V   **b)** V_p = 200 V   **c)** V_p = 80 V   **d)** V_p = 14 V   [3 marks each]

    a) ....

    b) ....

    c) ....

    d) ....

Medium questions: Use of $\frac{V_p}{V_s} = \frac{n_p}{n_s}$

Model example: A transformer has potential differences across the primary coil $V_p = 6$ V and across the secondary coil $V_p = 120$ V. If there are 100 turns on the primary coil ($n_p = 100$), calculate how many turns are in the secondary coil. [2 marks]

Write down equation:

$$\frac{V_p}{V_s} = \frac{n_p}{n_s}$$

Substitute variables into equation:

$$\frac{6}{120} = \frac{100}{n_s}$$ [1 mark]

Rearrange equation and calculate answer:

$n_s = 100 \div (6 \div 120) = 2000$ [1 mark]

Q3. A transformer has potential differences across the primary coil $V_p = 3$ V and across the secondary coil $V_s = 480$ V. Calculate how many turns are in the secondary coil, for each number of turns in the primary coil.

a) $n_p = 100$ **b)** $n_p = 20$ **c)** $n_p = 40$ **d)** $n_p = 150$ [2 marks each]

a) ....

b) ....

c) ....

d) ....

Q4. A transformer has potential differences across the primary coil $V_p = 200\ 000$ V and across the secondary coil $V_s = 40$ V. Calculate how many turns are in the primary coil, for each number of turns in the secondary coil.

a) $n_s = 10$ **b)** $n_s = 40$ **c)** $n_s = 80$ **d)** $n_s = 50$ [2 marks each]

a) ....

b) ....

c) ....

d) ....

Hard questions: Unit conversion needed.

Q5. A step-down transformer has potential differences of 230 V across the secondary coil and 420 kV across the primary coil. There are 4600 turns on the secondary coil.

a) Calculate how many turns there are on the primary coil. [3 marks]

....

b) There is a current of 13 A in the secondary coil. Calculate the power output. [3 marks]

....

[Total marks / 46]

Answers

Significant figures and standard form

Q1. **a)** 5×10^3

    b) 2.4×10^4

    c) 1.7×10^2

    d) 1.5×10^5

Q2. **a)** 430 000

    b) 320

    c) 94 000

    d) 430 000

Q3. **a)** 2.3×10^{-2}

    b) 3.2×10^{-3}

    c) 5.6×10^{-1}

    d) 8.73×10^{-3}

Q4.

Number	To three significant figures	To two significant figures	To one significant figure
4324	4320	4300	4000
8431	8430	8400	8000
0.274	0.274	0.27	0.3
4.308	4.31	4.3	4
0.00239	0.00239	0.0024	0.002

Mark scheme for Q5–10 (example for Q5)

Correct to two significant figures:

280 000 000   [1]

Correct conversion to standard form:

2.8×10^8 [1]

Q5. 2.8×10^8

Q6. 8.4×10^{-4}

Q7. 8.74×10^3

Q8. 2×10^{-2}

Q9. 4.4×10^5

Q10. 2.42×10^{-1}

Rearranging equations

Mark scheme for Q1–4 (example for Q1.a)

Correct substitution into equation:

$6 = 12 \times I$ [1]

Correct calculation of answer:

$I = 0.5$ [1]

Correct units:

amps (A)   [1]

Q1. **a)** 0.5 A

    b) 2 A

    c) 1.5 A

    d) 7.5 A

Q2. 50 V

Q3. **a)** 20 s

    b) 400 s

    c) 30 s

    d) 240 s

Q4. **a)** 17 m/s

    b) 14 m/s

    c) 16 m/s

    d) 12 m/s

Mean and uncertainties

Q1. 3.1 V

Q2. 6 paperclips

Q3. Anomaly = 11.5 A

    Mean = 13.0 A

Q4. Anomaly = 60.5 Ω

    Mean = 50.1 Ω

Mark scheme for Q5–6 (example for Q5)

Correct calculation of mean:

37.0 °C   [1]

Correct calculation of range:

1.0 °C   [1]

Correct calculation of uncertainty:

0.5 °C   [1]

Correct answer as mean ± uncertainty:

37.0 ± 0.5 °C

Q5. 37.0 ± 0.5 °C

Q6. 0.6 ± 0.1 kg

Mark scheme for Q7

Correct calculation of at least 2 resistances:

0.83, 0.87, 1.03, 0.83 Ω   [1]

Correct calculation of all 4 resistances: [1]

Correct identification of anomaly:

1.03 Ω   [1]

Correct calculation of mean:

0.84 Ω   [1]

Correct calculation of range:

0.04 Ω   [1]

Correct calculation of uncertainty:

0.02 Ω   [1]

Correct answer as mean ± uncertainty:

0.84 ± 0.02 Ω

Percentages and percentage changes

Q1. a) 63%

    b) 25%

    c) 4%

    d) 100%

Q2. a) 50%

    b) 1%

    c) 0.02%

    d) 8%

Mark scheme for Q3–Q8 (example for Q3.a)

Correct calculation of difference in speed:

60 − 30 = 30 m/s

Correct division of difference in speed by original speed:

$\frac{30}{30} = 1$ [1]

Correct percentage change:

1 × 100 = 100%   [1]

Q3. a) 100%

    b) 50%

    c) 200%

    d) 20%

Q4. 70%

Q5. 20% increase

Q6. 20% decrease

Q7. 5% increase

Q8. 25% increase

Gradients

Q1. a) 20

    b) 6

    c) −5

    d) 0.2

Mark scheme for Q2 (example for Q2.a)

Correct drawing of gradient triangle: [1]

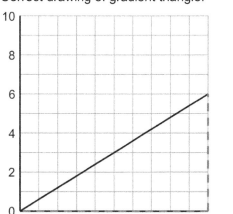

Correct calculation of gradient:

$\frac{6}{10} = 0.6$ [1]

Q2. a) 0.6

    b) 0.8

    c) 0.3

    d) −0.6

    e) −0.5

    f) −0.6

Mark scheme for Q3 (example for Q3.a)

Correct drawing of tangent.   [1]

Correct drawing of gradient triangle: [1]

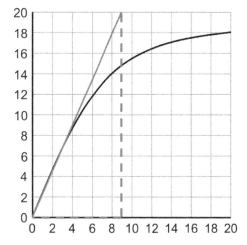

Correct calculation of gradient:

$\frac{20}{9} = 2.2$ (allow a range from 2 to 2.5)   [1]

Q3. a) 2–2.5

    b) 1.2–1.4

    c) 0.1–0.2

Answers

Magnification

Mark scheme for Q1–2 (example for Q1.a)

Correct substitution into equation:

$magnification = \frac{8}{0.5}$ [1]

Correct calculation of answer:

$magnification = \frac{8}{0.5} = 16$ [1]

Q1. a) 16

   b) 50

   c) 40

   d) 80

Q2. a) 15

   b) 75

   c) 45

   d) 8

Mark scheme for Q3–7 (example for Q3.a)

Correct unit conversion:

50 µm = 0.05 mm [1]

Correct substitution into equation:

$magnification = \frac{10}{0.05}$ [1]

Correct calculation of answer:

$magnification = \frac{10}{0.05} = 200$ [1]

Q3. a) 200

   b) 300

   c) 40

   d) 30

Q4. a) 80

   b) 320

   c) 200

   d) 240

Q5. 1.2 mm

Q6. 6 µm

Q7. 40 mm

Mark scheme for Q8

Correct calculation of the length of one palisade cell:

$\frac{240}{12} = 20$ µm [1]

Correct substitution into equation:

$50 = \frac{size\ of\ image}{20}$ [1]

Correct calculation of answer:

$size\ of\ image = 50 \times 20 = 1000$ [1]

Correct unit conversion:

1000 µm = 1 mm [1]

Q8. 1 mm

Culturing microorganisms

Mark scheme for Q1 (example for Q1.a)

Correct substitution into equation:

$cross\text{-}sectional\ area = \pi \times 1.5^2$ [1]

Correct calculation of answer:

$cross\text{-}sectional\ area = \pi \times 1.5^2 = 7.1$ [1]

Correct units:

cm^2 [1]

Q1. a) 7.1 cm^2

   b) 79 cm^2

   c) 50 cm^2

   d) 150 cm^2

Mark scheme for Q2–3 (example for Q2.a)

Correct calculation of number of divisions:

$\frac{30}{30} = 1$ division [1]

Correct substitution into equation:

$number\ of\ bacteria = 50 \times 2^1$ [1]

Correct calculation of answer:

$number\ of\ bacteria = 50 \times 2^1 = 100$ [1]

Q2. a) 100

   b) 3200

   c) 800

   d) 400

Q3. a) 320

   b) 80

   c) 160

   d) 1300

Mark scheme for Q4–6 (example for Q4)

Correct substitution into equation:

$6 = \pi \times r^2$ [1]

Correct calculation of answer:

$r = 1.4$ [1]

Correct unit conversion:

1.4 cm = 14 mm [1]

$r = 14$ mm [1]

Q4. 14 mm

Q5. 1.3×10^6

Q6. 11 mm

Surface area to volume ratio

Q1. a) 3.2 : 1

   b) 5 : 1

c) 4 : 1

d) 6.4 : 1

Mark scheme for Q2 (example for Q2.a)

Correct calculation of surface area of cube:

Each side = 1 × 1 = 1 cm^2

Six sides overall, so total surface area

= 6 × 1 = 6 cm^2 [1]

Correct calculation of volume of cube:

$Volume$ = 1 × 1 × 1 = 1 cm^3 [1]

Correct calculation of surface area to volume ratio:

$surface\ area \div volume$ = 6 ÷ 1 = 6

$surface\ area : volume\ ratio$ = 6 : 1 [1]

Q2. a) 6 : 1

 b) 3 : 1

 c) 2 : 1

 d) 1.5 : 1

Q3. Surface area to volume ratio decreases.

Mark scheme for Q4–5 (example for Q4)

Correct substitution into equation:

1.5 = 136 ÷ $volume$ [1]

Correct calculation of answer:

$volume$ = 136 ÷ 1.5 = 91 [1]

Correct units:

μm^3 [1]

Q4. 91 μm^3

Q5. 150 μm^2

Mark scheme for Q6.a

Correct substitution into equation:

$15 = \dfrac{10.5}{size\ of\ real\ object}$ [1]

Correct calculation of answer:

$size\ of\ real\ object = \dfrac{10.5}{15} = 0.7$ mm [1]

Mark scheme for Q6.b

Correct substitution into equation:

$25 = \dfrac{size\ of\ image}{0.7}$ [1]

Correct calculation of answer:

$size\ of\ image$ = 25 × 0.7 = 18 [1]

Correct units:

mm [1]

Mark scheme for Q6.c

Correct calculation of surface area of cube:

Each side = 0.7 × 0.7 = 0.49 mm^2

Six sides overall, so total surface area

= 6 × 0.49 = 2.94 mm^2 [1]

Correct calculation of volume of cube:

$Volume$ = 0.7 × 0.7 × 0.7 = 0.343 mm^3 [1]

Correct calculation of surface area to volume ratio:

$surface\ area \div volume$ = 2.94 ÷ 0.343 = 8.6

$surface\ area : volume\ ratio$ = 8.6 : 1 [1]

Q6. a) 0.7 mm

 b) 18 mm

 c) 8.6 : 1

Rates

Mark scheme for Q1–3 (example for Q1.a)

Correct substitution into equation:

$breaths\ per\ minute = \dfrac{280}{7}$ [1]

Correct calculation of answer:

$breaths\ per\ minute = \dfrac{280}{7}$

= 40 breaths per minute [1]

Q1. a) 40 breaths per minute

 b) 70 breaths per minute

 c) 14 breaths per minute

 d) 19 breaths per minute

Q2. a) 5 ml/min

 b) 4 ml/min

 c) 3 ml/min

 d) 7.5 ml/min

Q3. a) 360

 b) 600

 c) 1800

 d) 840

Mark scheme for Q4 (example for Q4.a)

Correct unit conversion:

1 hour = 60 minutes [1]

Correct substitution into equation:

$breaths\ per\ minute = \dfrac{2000}{60}$ [1]

Correct calculation of answer:

$breaths\ per\ minute = \dfrac{2000}{60}$

= 33 breaths per minute [1]

Q4. a) 33 breaths per minute

 b) 20 breaths per minute

 c) 14 breaths per minute

 d) 11 breaths per minute

Mark scheme for Q5 (example for Q5.a)

Correct calculation of light intensity:

$light\ intensity \propto \frac{1}{distance^2}$

Light intensity is $\frac{1}{4}$ of the original intensity. [1]

Correct calculation of bubbles of oxygen:

$\frac{1}{4} \times 36 = 9$ bubbles of oxygen per minute [1]

Q5. a) 9 bubbles of oxygen per minute

    b) 4 bubbles of oxygen per minute

Genetic diagrams and probabilities

Mark scheme for Q1 (example for Q1.a)

Correct identification of the brown hair genotypes:

BB and Bb [1]

Correct calculation of probability:

Four potential genotypes. Two out of four are Bb.

$\frac{1}{2}$ probability the child will have brown hair. [1]

Q1. a) $\frac{1}{2}$

    b) 1

    c) 0

    d) $\frac{3}{4}$

Mark scheme for Q2 (example for Q2.a)

Correct identification of the short genotype:

tt [1]

Correct calculation of probability:

Four potential genotypes. Two out of four are tt.

$\frac{1}{2}$ probability the offspring will be short. [1]

Correct calculation of percentage probability:

$\frac{1}{2} \times 100 = 50\%$ [1]

Q2. a) 50%

    b) 0%

Mark scheme for Q3–4 (example for Q3)

Correct genetic diagram (other letters allowed): [1]

Correct identification of the carrier genotype:

Ff [1]

Correct calculation of probability:

Four potential genotypes. Two out of four are Ff.

$\frac{1}{2}$ probability the offspring will be a carrier. [1]

Correct calculation of percentage probability:

$\frac{1}{2} \times 100 = 50\%$ [1]

Q3. 50%

Q4. 25%

	P	p
P	PP	Pp
p	Pp	pp

Using quadrats

Q1. a) 24 000

    b) 27 000

    c) 12 000

    d) 42 000

Q2. a) 80 000

    b) 50 000

    c) 150 000

    d) 36 000

Q3. a) 15

    b) 60 000

Mark scheme for Q3.c

Correct calculation of number of daisies in each quadrat:

$15 \times 2 = 30$ [1]

Correct calculation of total number of daisies:

$30 \times 10\ 000 = 300\ 000$ [1]

Answer correctly written in standard form:

$300\ 000 = 3 \times 10^5$ [1]

Q3. c) 3×10^5

Mark scheme for Q4.a

Correct calculation of area:

$16 + 40 = 56$ m^2 [1]

Correct calculation of mean number of cornflowers:

$(8 + 4 + 12 + 20 + 0) \div 5 = 8.8$ [1]

Correct calculation of total number of cornflowers:

$8.8 \times 56 = 490$ [1]

Mark scheme for Q4.b

Correct calculation of new area:

$56 - 4 = 52$ m^2 [1]

Correct calculation of total number of cornflowers:

$8.8 \times 52 = 460$ [1]

Q4. a) 490

    b) 460

Efficiency of biomass transfer

Mark scheme for Q1–3 (example for Q1.a)

Correct substitution into equation:

$efficiency\ of\ biomass\ transfer = \frac{200}{2200} \times 100$ [1]

Correct calculation of answer:

$\frac{200}{2200} \times 100 = 9.1\%$ [1]

Q1. a) 9.1%

    b) 11%

    c) 10%

    d) 8.3%

Q2. a) 10%

    b) 9.3%

    c) 13%

    d) 8%

Q3. a) 8 g

    b) 6.4 g

    c) 9.6 g

    d) 2.4 g

Mark scheme for Q4 (example for Q4.a)

Correct unit conversion:

400 g = 0.4 kg [1]

Correct substitution into equation:

$efficiency\ of\ biomass\ transfer = \frac{0.4}{5} \times 100$ [1]

Correct calculation of answer:

$\frac{0.4}{5} \times 100 = 8\%$ [1]

Q4. a) 8%

    b) 7.6%

    c) 11%

    d) 9%

Mark scheme for Q5

Correct calculation of increase in mass of elephant:

160 − 120 − 20 = 20 kg [1]

Correct substitution into equation:

$efficiency\ of\ biomass\ transfer = \frac{20}{160} \times 100$ [1]

Correct calculation of answer:

$\frac{20}{160} \times 100 = 12.5\%$ [1]

Q5. 13%

Mark scheme for Q6

Correct unit conversion:

0.2 kg = 200 g [1]

Correct calculation of increase in mass of mongoose:

200 − 40 − 145 = 15 g [1]

Correct substitution into equation:

$efficiency\ of\ biomass\ transfer = \frac{15}{200} \times 100$ [1]

Correct calculation of answer:

$\frac{15}{200} \times 100 = 7.5\%$ [1]

Q6. 7.5%

Mark scheme for Q7

Correct unit conversion:

0.15 kg = 150 g [1]

Correct substitution into equation:

$8 = \frac{biomass\ of\ higher\ trophic\ level}{150} \times 100$ [1]

Correct calculation of biomass of higher trophic level:

$biomass\ of\ higher\ trophic\ level = \frac{(8 \times 150)}{100} = 12\ g$ [1]

Correct calculation of mass used in respiration:

150 − 90 − 12 = 48 g [1]

Q7. 48 g

Relative formula and atomic mass

Mark scheme for Q1 (example for Q1.a)

Correct substitution of atomic masses:

16 × 2 [1]

Correct calculation of relative formula mass:

16 × 2 = 32 [1]

Q1. a) 32

    b) 28

    c) 36.5

    d) 101

    e) 160

    f) 98

    g) 74

    h) 331

Mark scheme for Q2–8 (example for Q2)

Correct substitution into equation:

$relative\ atomic\ mass = \frac{(35 \times 76) + (37 \times 24)}{100}$ [1]

Correct calculation of relative formula mass:

$\frac{(35 \times 76) + (37 \times 24)}{100} = 35.5$ [1]

Q2. 35.5

Q3. 10.8

Q4. 63.6

Q5. 69.4

Q6. 80.0

Q7. 20.2

Q8. 24.3

Bulk and surface properties of matter

Mark scheme for Q1–2 (example for Q1.a)

Correct substitution into equation:

surface area ÷ *volume* = 8000 ÷ 80 000   [1]

Correct calculation of answer:

8000 ÷ 80 000 = 0.1   [1]

Correct units:

nm^{-1} [1]

Q1. a) $0.1 \ nm^{-1}$

     b) $0.08 \ nm^{-1}$

     c) $0.057 \ nm^{-1}$

     d) $0.13 \ nm^{-1}$

Q2. a) $2500 \ nm^2$

     b) $2000 \ nm^2$

     c) $10 \ 000 \ nm^2$

     d) $11 \ 000 \ nm^2$

Mark scheme for Q3 (example for Q3.a)

Correct calculation of surface area:

six sides overall, so total surface area

$= 6 \times 1 \times 1 = 6 \ nm^2$ [1]

Correct calculation of volume

volume $= 1 \times 1 \times 1 = 1 \ nm^3$ [1]

Correct calculation of answer:

6 ÷ 1 = 6   [1]

Correct units:

nm^{-1} [1]

Q3. a) $6 \ nm^{-1}$

     b) $1.2 \ nm^{-1}$

     c) $0.4 \ nm^{-1}$

     d) $1.5 \ nm^{-1}$

Percentage mass

Mark scheme for Q1–6 (example for Q1)

Correct substitution into equation:

$\% \ mass = \frac{12}{28} \times 100$ [1]

Correct calculation:

$\% \ mass = \frac{12}{28} \times 100 = 43\%$ [1]

Q1. 43%

Q2. 2.7%

Q3. 48%

Q4. 40%

Q5. 39%

Q6. 56%

Mark scheme for Q7–15 (example for Q7)

Correct calculation of formula mass:

12 + (2 × 16) = 44   [1]

Correct substitution into equation:

$\% \ mass = \frac{12}{44} \times 100$ [1]

Correct calculation:

$\% \ mass = \frac{12}{44} \times 100 = 27\%$ [1]

Q7. 27%

Q8. 50%

Q9. 7.7%

Q10. 40%

Q11. 29%

Q12. 16%

Q13. 19%

Q14. 36%

Q15. 21%

Moles

Q1. a) 0.5 mol

     b) 20 mol

     c) 3 mol

     d) 0.05 mol

Q2. a) 168 g

     b) 24 g

     c) 23.8 g

     d) 3.25 g

Mark scheme for Q3 (example for Q3.a)

Correct relative formula mass calculation:

$relative \ formula \ mass = \frac{mass}{number \ of \ moles}$

$\frac{130}{2} = 75 \ g/mol$ [1]

Correct identification of the element:

arsenic (as it has a relative atomic mass of 75)   [1]

Q3. a) $M_r = 75$; arsenic

     b) $M_r = 56$; iron

     c) $M_r = 96$; molybdenum

     d) $M_r = 39$; potassium

Mark scheme for Q4–5 (example for Q4.a)

Correct relative formula mass calculation:

(2 × 16) = 32 g/mol   [1]

Correct substitution and calculation:

$relative \ formula \ mass = \frac{mass}{number \ of \ moles}$

$\frac{80}{32}$ = 2.5 mol [1]

Q4. a) 2.5 mol

b) 0.1 mol

c) 2 mol

d) 0.5 mol

Q5. a) 10.8 g

b) 116 g

c) 214 g

d) 3 g

Mark scheme for Q6–8 (example for Q6)

Correct unit conversion:

4.59 kg = 4590 g [1]

Correct relative formula mass calculation:

$(2 \times 27) + (3 \times 16)$ = 102 g/mol [1]

Correct substitution and calculation:

$relative\ formula\ mass = \frac{mass}{number\ of\ moles}$

$\frac{4290}{102}$ = 45 mol [1]

Q6. 45 mol

Q7. 2×10^{-5} mol

Q8. 30 mol

Balancing symbol equations

Q1. a) $2Na + Cl_2 \rightarrow 2NaCl$

b) $Zn + 2HCl \rightarrow ZnCl_2 + H_2$

c) $2Mg + O_2 \rightarrow 2MgO$

d) $4Al + 3O_2 \rightarrow 2Al_2O_3$

e) $2K + 2H_2O \rightarrow 2KOH + H_2$

f) $2NaOH + H_2SO_4 \rightarrow 2H_2O + Na_2SO_4$

Mark scheme for Q2

Correct left-hand side of equation. [1]

Correct right-hand side of equation. [1]

Q2. a) $2H_2O_2 \rightarrow O_2 + 2H_2O$

b) $N_2 + 3H_2 \rightarrow 2NH_3$

c) $2Na + H_2SO_4 \rightarrow Na_2SO_4 + H_2$

d) $6CO_2 + 6H_2O \rightarrow C_6H_{12}O_6 + 6O_2$

e) $CuCO_3 \rightarrow CuO + CO_2$

f) $2C_6H_6 + 15O_2 \rightarrow 12CO_2 + 6H_2O$

Mark scheme for Q3–4 (example for Q3)

Correct calculation of relative formula masses:

Zn: 65

$CuSO_4$: 63.5 + 32 + (4 × 16) = 159.5

$ZnSO_4$: 65 + 32 + (4 × 16) = 161

Cu: 63.5 [1]

Correct calculation of number of moles of each substance:

Zn: 2 mol, $CuSO_4$: 2 mol, $ZnSO_4$: 2 mol, Cu: 2 mol [1]

Correct balanced symbol equation:

$Zn + CuSO_4 \rightarrow Cu + ZnSO_4$ [1]

Q3. $Zn + CuSO_4 \rightarrow Cu + ZnSO_4$

Q4. $2K + 2H_2O \rightarrow 2KOH + H_2$

Concentration of solutions

Mark scheme for Q1–4 (example for Q1.a)

Correct substitution into equation:

$concentration = \frac{20}{8}$ [1]

Correct calculation:

$concentration = \frac{20}{8} = 2.5$ [1]

Correct units:

g/dm^3 [1]

Q1. a) 2.5 g/dm^3

b) 0.5 g/dm^3

c) 40 g/dm^3

d) 6.7 g/dm^3

Q2. a) 0.2 g/dm^3

b) 0.063 g/dm^3

c) 0.042 g/dm^3

d) 2.5 g/dm^3

Q3. a) 0.05 g

b) 0.02 g

c) 0.075 g

d) 0.004 g

Q4. a) 1.3 dm^3

b) 0.63 dm^3

c) 0.25 dm^3

d) 1.9 dm^3

Q5. a) 0.03 mol

b) 0.015 mol

c) 0.2 mol/dm^3

Mark scheme for Q5.c

Correct substitution into equation:

$concentration = \frac{0.015}{0.075}$ [1]

Correct calculation:

$concentration = \frac{0.015}{0.075} = 0.02$ [1]

Correct units:

mol/dm^3 [1]

Q6. a) 0.0004 mol

    b) 0.0004 mol

    c) 0.01 mol/dm^3

Mark scheme for Q6.c

Correct substitution into equation:

$0.04 = \dfrac{0.0004}{volume\ of\ solution}$ [1]

Correct calculation:

$volume\ of\ solution = \dfrac{0.0004}{0.04} = 0.01$ [1]

Correct units:

mol/dm^3 [1]

Volumes of gases

Mark scheme for Q1 (example for Q1.a)

Correct substitution into equation:

$volume\ of\ gas\ (\mathrm{dm}^3) = 24 \times \dfrac{14}{28}$ [1]

Correct calculation:

$volume\ of\ gas\ (\mathrm{dm}^3) = 24 \times \dfrac{14}{28} = 12$ [1]

Correct units:

dm^3 [1]

Q1. a) 12 dm^3

    b) 18 dm^3

    c) 34 dm^3

    d) 4.3 dm^3

Mark scheme for Q2–5 (example for Q2)

Correct calculation of M_r of gas:

$M_r = 2 \times 1 = 2$ [1]

Correct substitution into equation:

$volume\ of\ gas\ (\mathrm{dm}^3) = 24 \times \dfrac{40}{2}$ [1]

Correct calculation:

$volume\ of\ gas\ (\mathrm{dm}^3) = 24 \times \dfrac{40}{2} = 480$ [1]

Correct units:

dm^3 [1]

Q2. 480 dm^3

Q3. 19 dm^3

Q4. 96 dm^3

Q5. 91 dm^3

Mark scheme for Q6–8 (example for Q6)

Correct unit conversion:

200 cm^3 = 0.2 dm^3 [1]

Correct calculation of M_r of gas:

$M_r = 2 \times 35.5 = 71$ [1]

Correct substitution into equation:

$0.2 = 24 \times \dfrac{mass\ of\ gas}{71}$ [1]

Correct calculation:

$mass\ of\ gas = 0.2 \times \dfrac{71}{24} = 0.59$ [1]

Correct units:

g [1]

Q6. 0.59 g

Q7. 0.84 g

Q8. 11 g

Percentage yield

Mark scheme for Q1–2 (example for Q1.a)

Correct substitution into equation:

$\%\ yield = \dfrac{1200}{2000} \times 100$ [1]

Correct calculation:

$\%\ yield = \dfrac{1200}{2000} \times 100 = 60\%$ [1]

Q1. a) 60%

    b) 75%

    c) 5%

    d) 25%

Q2. a) 15%

    b) 55%

    c) 75%

    d) 5%

Mark scheme for Q3–4 (example for Q3.a)

Correct substitution into equation:

$60 = \dfrac{mass\ of\ product\ actually\ made}{2000} \times 100$ [1]

Correct calculation:

$mass\ of\ product\ actually\ made$

$= 60 \times \dfrac{2000}{100} = 1200$ [1]

Correct units:

g [1]

Q3. a) 1200 g

    b) 300 g

    c) 480 g

    d) 24 g

Q4. a) 2000 g

    b) 125 g

    c) 150 g

    d) 1100 g

Mark scheme for Q5.a and Q6.a (example for Q5.a)

Correct calculation of number of moles of zinc:

$moles = \frac{6.5}{65} = 0.1$ mol [1]

Correct statement of number of moles of $ZnSO_4$:

0.1 mol [1]

Correct calculation of M_r of $ZnSO_4$:

65 + 32 + 64 = 161 [1]

Correct mass of $ZnSO_4$:

$0.1 = \frac{mass}{161}$, mass = 16 g [1]

Mark scheme for Q5.b and Q6.b (example for Q5.b)

Correct substitution into equation:

$\% yield = \frac{6}{16.1} \times 100$ [1]

Correct calculation:

$\% yield = \frac{6}{16.1} \times 100 = 37\%$ [1]

Q5. a) 16 g

   b) 37%

Q6. a) 320 g

   b) 80%

Atom economy

Mark scheme for Q1–2 (example for Q1.a)

Correct substitution into equation:

$percentage\ atom\ economy = \frac{12}{24} \times 100$ [1]

Correct calculation:

$percentage\ atom\ economy = \frac{12}{24} \times 100 = 50\%$ [1]

Q1. a) 50%

   b) 40%

   c) 33%

   d) 30%

Q2. a) 80%

   b) 40%

   c) 25%

   d) 91%

Mark scheme for Q3 (example for Q3.a)

Correct substitution into equation:

$80 = \frac{80}{sum\ of\ relative\ formula\ masses\ of\ all\ reactants} \times 100$ [1]

Correct calculation:

$sum\ of\ relative\ formula\ masses\ of\ all\ reactants$

$= 80 \times \frac{100}{80} = 100$ [1]

Correct units:

g [1]

Q3. a) 100 g

   b) 150 g

   c) 19 g

   d) 31 g

Mark scheme for Q4–7 (example for Q4)

Correct calculation of M_r of desired product:

$2 \times (2 \times 1) = 4$ [1]

Correct calculation of M_r of all reactants:

$2 \times [(2 \times 1) + 16)] = 36$ [1]

Correct substitution into equation:

$percentage\ atom\ economy = \frac{4}{36} \times 100$ [1]

Correct calculation:

$percentage\ atom\ economy = \frac{4}{36} \times 100 = 11\%$ [1]

Q4. 11%

Q5. 56%

Q6. 26%

Q7. 51%

Bond energy calculations

Q1. a) 1598 kJ

   b) 1652 kJ

   c) 1173 kJ

   d) 1384 kJ

Q2. a) 1566 kJ

   b) 2266 kJ

   c) 2826 kJ

   d) 3235 kJ

Mark scheme for Q3–4 (example for Q3)

Correct calculation of total bond energy of reactants:

498 kJ [1]

Correct calculation of total bond energy or products:

2 × 799 = 1598 kJ [1]

Correct calculation of total energy change:

498 − 1598 = −1100 kJ [1]

Correct exothermic/endothermic indication:

Energy change is negative; exothermic. [1]

Q3. −1100 kJ, exothermic

Q4. −806 kJ, exothermic

Rates of reaction

Mark scheme for Q1–4 (example for Q1.a)

Correct substitution into equation:

$mean\ rate\ of\ reaction = \frac{400}{8}$ [1]

Correct calculation:

$mean\ rate\ of\ reaction = \frac{400}{8} = 50$ [1]

Correct units:

g/s   [1]

Q1. a) 50 g/s

    b) 1.6 g/s

    c) 1.3 g/s

    d) 16 g/s

Q2. a) 12 cm^3/s

    b) 2.4 cm^3/s

    c) 0.25 cm^3/s

    d) 15 cm^3/s

Q3. a) 200 mol

    b) 500 mol

    c) 4 mol

    d) 60 mol

Q4. a) 2.5 s

    b) 8 s

    c) 0.6 s

    d) 36 s

Mark scheme for Q5.b and Q5.c (example for Q5.b)

Correct drawing of tangent:

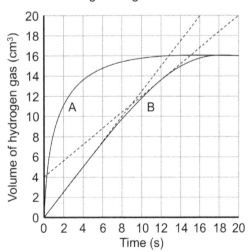

As shown in diagram.   [1]

Correct drawing of gradient triangle.   [1]

Correct calculation of gradient:

20 ÷ 16 = 1.3   [1]

Correct units:

cm^3/s   [1]

Q5. a) Graph A

    b) 1.3 cm^3/s (range of 1.1 – 1.5 cm^3/s allowed)

    c) 0.8 cm^3/s (range of 0.6 – 1.0 cm^3/s allowed)

Chromatography

Mark scheme for Q1–2 (example for Q1.a)

Correct substitution into equation:

$R_f = \frac{6}{12}$ [1]

Correct calculation:

$R_f = \frac{6}{12} = 0.5$ [1]

Q1. a) 0.5

    b) 0.17

    c) 0.83

    d) 0.13

Q2. a) 0.5

    b) 0.38

    c) 0.33

    d) 0.75

Mark scheme for Q3–4 (example for Q3.a)

Correct substitution into equation:

$0.7 = \frac{distance\ moved\ by\ substance}{4}$ [1]

Correct calculation:

$distance\ moved\ by\ substance = 0.7 \times 4 = 2.8$ [1]

Correct units:

cm   [1]

Q3. a) 2.8 cm

    b) 3.5 cm

    c) 4.9 cm

    d) 2.1 cm

Q4. a) 3.3 cm

    b) 5 cm

    c) 15 cm

    d) 2.5 cm

Mark scheme for Q5.a–Q5.c (example for Q5.a)

Correct measurement of distance travelled by the solvent:

4.1 cm   [1]

Correct measurement of distance travelled by the colour:

3.1 cm   [1]

Correct calculation:

$R_f = \frac{3.1}{4.1} = 0.76$ (allow a range of 0.74 – 0.78) [1]

Q5. a) 0.76 (allow a range of 0.74 – 0.78)

    b) 0.39 (allow a range of 0.37 – 0.41)

    c) 0.66 (allow a range of 0.64 – 0.68)

    c) Colours 1 and 2.

Kinetic energy

Mark scheme for Q1–2 (example for Q1.a)

Correct substitution into equation:

$E_k = \frac{1}{2} \times 4 \times 10^2$ [1]

Correct calculation of answer:

$E_k = \frac{1}{2} \times 4 \times 10^2 = 200$ [1]

Correct units:

joules (J) [1]

Q1. a) 200 J

    b) 8 J

    c) 0.5 J

    d) 2 J

Q2. a) 10 000 J

    b) 160 000 J

    c) 2500 J

    d) 400 J

Mark scheme for Q3–8 (example for Q3.a)

Correct unit conversion:

$m = 56$ g $= 0.056$ kg [1]

Correct substitution into equation:

$v = \sqrt{\frac{0.28}{0.5 \times 0.056}}$ [1]

Correct calculation of answer:

$v = \sqrt{\frac{0.28}{0.5 \times 0.056}} = 3.2$ [1]

Correct units:

metres per second (m/s) [1]

Q3. a) 3.2 m/s

    b) 13.0 m/s

    c) 1.6 m/s

    d) 4.7 m/s

Q4. a) 2.5 kg

    b) 1 kg

    c) 1.5 kg

    d) 4 kg

Q5. 80 m/s

Q6. 180 kg

Q7. 690 J

Q8. 40 m/s

Mark scheme for Q9

Correct substitution into equation:

$m = \frac{2 \times 500}{5^2}$ [1]

Correct calculation of total mass:

$m = \frac{2 \times 500}{5^2} = 40$ kg [1]

Correct subtraction of 2 kg skateboard mass:

$40 - 2 = 38$ kg [1]

Correct units:

kilograms (kg) [1]

Q9. 38 kg

Elastic potential energy

Mark scheme for Q1–2 (example for Q1.a)

Correct substitution into equation:

$E_e = \frac{1}{2} \times 40 \times 2^2$ [1]

Correct calculation of answer:

$E_e = \frac{1}{2} \times 40 \times 0.2^2 = 80$ [1]

Correct units:

joules (J) [1]

Q1. a) 80 J

    b) 45 J

    c) 1.8 J

    d) 0.2 J

Q2. a) 0.36 J

    b) 0.023 J

    c) 11 J

    d) 81 J

Mark scheme for Q3–6 (example for Q3.b): 3 marks if unit conversion not required, 4 marks if there is a unit conversion

Correct unit conversion:

$E_e = 0.8$ kJ $= 800$ J [1]

Correct substitution into equation:

$e = \sqrt{\frac{200}{0.5 \times 800}}$ [1]

Correct calculation of answer:

$e = \sqrt{\frac{200}{0.5 \times 800}} = 2$ [1]

Correct units:

metres (m) [1]

Q3. a) 1 m

    b) 2 m

c) 0.5 m

d) 3 m

Q4. a) 320 N/m

   b) 13 000 N/m

   c) 3100 N/m

   d) 50 000 N/m

Q5. 0.01 m

Q6. 120 N/m

Mark scheme for Q7 (example for Q7.b)

Correct unit conversion:

m = 250 g = 0.25 kg [1]

Correct substitution into equation:

$v = \sqrt{\dfrac{7.5}{0.5 \times 0.25}}$ [1]

Correct calculation of answer:

$v = \sqrt{\dfrac{7.5}{0.5 \times 0.25}} = 7.7$ [1]

Correct units:

metres per second (m/s) [1]

Q7. a) 7.5 J

   b) 7.7 m/s

Gravitational potential energy

Mark scheme for Q1–5 (example for Q1.a)

Correct substitution into equation:

E_p = 0.4 × 9.8 × 1 [1]

Correct calculation of answer:

E_p = 0.4 × 9.8 × 1 = 3.9 [1]

Correct units:

joules (J) [1]

Q1. a) 3.9 J

   b) 3.1 J

   c) 5.9 J

   d) 9.8 J

Q2. a) 10 m

   b) 2 m

   c) 8 m

   d) 12 m

Q3. a) 0.1 kg

   b) 0.5 kg

   c) 0.4 kg

   d) 0.7 kg

Q4. 31 m

Q5. 2 m

Mark scheme for Q6–Q7 (example for Q6)

Correct unit conversion:

h = 8 km = 8000 m [1]

Correct substitution into equation:

$m = \dfrac{3.2 \times 10^9}{9.8 \times 8000}$ [1]

Correct calculation of answer:

$m = \dfrac{3.2 \times 10^9}{9.8 \times 8000} = 41\,000$ [1]

Correct units:

kilograms (kg) [1]

Q6. 41 000 kg

Q7. a) 63 J

   b) 13 m/s

Mark scheme for Q8

Correct unit conversion:

m = 56 g = 0.056 kg [1]

Correct substitution into equation:

E_p = 0.056 × 9.8 × 6 [1]

Correct calculation of answer:

E_p = 0.056 × 9.8 × 6 = 3.3 J [1]

Correct substitution into equation:

$v = \sqrt{\dfrac{3.3}{0.5 \times 0.056}}$ [1]

Correct calculation of answer:

$v = \sqrt{\dfrac{3.3}{0.5 \times 0.056}} = 11$ [1]

Correct units:

metres per second (m/s) [1]

Q8. 11 m/s

Specific heat capacity

Mark scheme for Q1–7 (example for Q1.a)

Correct substitution into equation:

ΔE = 0.4 × 4200 × 10 [1]

Correct calculation of answer:

ΔE = 0.4 × 4200 × 10 = 17 000 [1]

Correct units:

joules (J) [1]

Q1. a) 17 000 J

   b) 42 000 J

   c) 100 000 J

   d) 3400 J

Q2. a) 900 J

   b) 2300 J

   c) 360 000 J

d) 1 100 000 J

Q3. 170 000 J

Q4. a) 210 °C

  b) 26 °C

  c) 1.0 °C

  d) 0.82 °C

Q5. a) 0.56 kg

  b) 0.37 kg

  c) 0.93 kg

  d) 1.4 kg

Q6. 120 kg

Q7. 3.3 °C

Mark scheme for Q8

Correct unit conversion:

m = 800 g = 0.8 kg [1]

Correct substitution into equation:

$\Delta\theta = \frac{1.5 \times 10^4}{0.8 \times 900}$ [1]

Correct calculation of temperature change:

$\Delta\theta = \frac{1.5 \times 10^4}{0.8 \times 900} = 21$ °C [1]

Correct calculation of final temperature:

21 + 20 = 41 °C [1]

Q8. 41 °C

Power

Mark scheme for Q1–4 (example for Q1.a)

Correct substitution into equation:

$P = \frac{100\ 000}{50}$ [1]

Correct calculation of answer:

$P = \frac{100\ 000}{50} = 2000$ [1]

Correct units:

watts (W) [1]

Q1. a) 2000 W

  b) 2500 W

  c) 1000 W

  d) 910 W

Q2. a) 5 W

  b) 13 W

  c) 40 W

  d) 16 W

Q3. 80 W

Q4. a) 500 s

  b) 2000 s

  c) 400 s

  d) 10 000 s

Mark scheme for Q5–7 (example for Q5.a)

Correct unit conversion:

P = 3 kW = 3000 W [1]

Correct unit conversion:

t = 2 minutes = 120 s [1]

Correct substitution into equation:

E = 3000 × 120 [1]

Correct calculation of answer:

E = 3000 × 120 = 360 000 [1]

Correct units:

joules (J) [1]

Q5. a) 360 000 J

  b) 900 000 J

  c) 3 200 000 J

  d) 11 000 000 J

Q6. 8 600 000 J

Q7. 190 000 J

Mark scheme for Q8

Correct substitution into equation:

E_p = 500 × 9.8 × 20 [1]

Correct calculation of answer:

E_p = 500 × 9.8 × 20 = 98 000 J [1]

Correct substitution into equation:

$P = \frac{98\ 000}{30} = 3300$ [1]

Correct units:

watts (W) [1]

Q8. 3300 W

Efficiency

Mark scheme for Q1–2 (example for Q1.a)

Correct substitution into equation:

$Efficiency = \frac{3000}{4000}$ [1]

Correct calculation of answer:

$Efficiency = \frac{3000}{4000} = 0.75$ [1]

Q1. a) 0.75

  b) 0.5

  c) 0.9

  d) 0.95

Q2. a) 0.5

  b) 0.91

c) 0.1

d) 0.33

Q8. 0.75

Mark scheme for Q3–6 (example for Q3.a)

Correct substitution into equation:

Useful output energy = 200 000 × 0.6 [1]

Correct calculation of answer:

200 000 × 0.6 = 120 000 [1]

Correct units:

joules (J) [1]

Q3. a) 120 000 J

b) 160 000 J

c) 30 000 J

d) 110 000 J

Q4. a) 100 000 W

b) 250 000 W

c) 200 000 W

d) 130 000 W

Q5. 10 W

Q6. 1300 MW

Mark scheme for Q7

Correct substitution into equation:

$E_p = 200 × 9.8 × 40$ [1]

Correct calculation of answer:

$E_p = 200 × 9.8 × 40 = 78\ 400$ J [1]

Correct unit conversion:

156.8 kJ = 156 800 J [1]

Correct substitution into equation:

$Efficiency = \frac{78\ 400}{156\ 800}$ [1]

Correct calculation of answer:

$Efficiency = \frac{78\ 400}{156\ 800} = 0.5$ [1]

Q7. 0.5

Mark scheme for Q8

Correct unit conversion:

10 cm = 0.1 m [1]

Correct substitution into equation:

$E_e = 0.5 × 80 × 0.1^2$ [1]

Correct calculation of answer:

$E_e = 0.5 × 80 × 0.1^2 = 0.4$ J [1]

Correct substitution into equation:

$Efficiency = \frac{0.3}{0.4}$ [1]

Correct calculation of answer:

$Efficiency = \frac{0.3}{0.4} = 0.75$ [1]

Charge flow

Mark scheme for Q1–2 (example for Q1.a)

Correct substitution into equation:

$Q = 6 × 90$ [1]

Correct calculation of answer:

$Q = 6 × 90 = 540$ [1]

Correct units:

coulombs (C) [1]

Q1. a) 540 C

b) 900 C

c) 360 C

d) 2400 C

Q2. a) 250 C

b) 900 C

c) 1400 C

d) 300 C

Mark scheme for Q3–5 (example for Q3.a)

Correct unit conversion:

30 minutes = 1800 s [1]

Correct substitution into equation:

$I = \frac{18\ 000}{1800}$ [1]

Correct calculation of answer:

$I = \frac{18\ 000}{1800} = 10$ [1]

Correct units:

amps (A) [1]

Q3. a) 10 A

b) 4 A

c) 5 A

d) 1.3 A

Q4. a) 1.5 A

b) 0.5 A

c) 0.38 A

d) 0.21 A

Q5. a) 1000 s

b) 750 s

c) 2500 s

d) 4200 s

Mark scheme for Q6.a

Correct substitution into equation:

$Q = (3.2 × 10^{-20}) × 5$ [1]

Correct calculation of answer:

$Q = (3.2 \times 10^{-20}) \times 5 = 1.6 \times 10^{-19}$ [1]

Correct units:

coulombs (C) [1]

Q6. a) 1.6×10^{-19} C

Mark scheme for Q6.b–c (example for Q6.b)

Correct unit conversion:

$t = 2$ ms $= 0.002$ s [1]

Correct substitution into equation:

$I = \frac{5 \times 10^{16} \times 1.6 \times 10^{-19}}{0.002}$ [1]

Correct calculation of answer:

$I = \frac{5 \times 10^{16} \times 1.6 \times 10^{-19}}{0.002} = 4$ [1]

Correct units:

amps (A) [1]

Q6. b) 4 A

   c) 5.3×10^{-7} A

Potential difference, current and resistance

Mark scheme for Q1–2 (example for Q1.a)

Correct substitution into equation:

$V = 0.5 \times 20$ [1]

Correct calculation of answer:

$V = 0.5 \times 20 = 10$ [1]

Correct units:

volts (V) [1]

Q1. a) 10 V

   b) 16 V

   c) 80 V

   d) 300 V

Q2. a) 100 V

   b) 20 V

   c) 5 V

   d) 75 V

Mark scheme for Q3–5 (example for Q3.a)

Correct unit conversion:

$I = 200$ mA $= 0.2$ A [1]

Correct substitution into equation:

$R = \frac{12}{0.2}$ [1]

Correct calculation of answer:

$R = \frac{12}{0.2} = 60\ \Omega$ [1]

Correct units:

ohms (Ω) [1]

Q3. a) 60 Ω

   b) 240 Ω

   c) 3000 Ω

   d) 400 Ω

Q4. a) 0.06 A

   b) 0.012 A

   c) 0.02 A

   d) 0.001 A

Q5. 500 A

Mark scheme for Q6

Correct unit conversion:

$V = 5$ MV $= 5\ 000\ 000$ V [1]

Correct unit conversion:

$R = 25$ kΩ $= 25\ 000\ \Omega$ [1]

Correct substitution into equation:

$I = \frac{5\ 000\ 000}{25\ 000}$ [1]

Correct calculation of temperature change:

$I = \frac{5\ 000\ 000}{25\ 000} = 200$ [1]

Correct units:

amps (A) [1]

Q6. 200 A

Mark scheme for Q7

Correct substitution into equation:

$I = \frac{130}{10}$ [1]

Correct calculation of answer:

$I = \frac{130}{10} = 13$ A [1]

Correct substitution into equation:

$V = 13 \times 5$ [1]

Correct calculation of answer:

$V = 13 \times 5 = 65$ [1]

Correct units:

volts (V) [1]

Q7. 65 V

Mark scheme for Q8

Correct substitution into equation:

$I = \frac{12}{60}$ [1]

Correct calculation of answer:

$I = \frac{12}{60} = 0.2$ A [1]

Correct unit conversion

$t = 2$ minutes $= 120$ s [1]

Correct substitution into equation:

$Q = 0.2 \times 120$ [1]

Correct calculation of answer:

$Q = 0.2 \times 120 = 24$ [1]

Correct units:

coulombs (C) [1]

Q8. 24 C

Series and parallel circuits

Q1. a) 15 Ω

    b) 225 Ω

    c) 2400 Ω

    d) 0.5 Ω

Mark scheme for Q2.b

Correct substitution into equation:

$I = \frac{12}{60}$ [1]

Correct calculation of answer:

$I = \frac{12}{60} = 0.2$ [1]

Correct units:

amps (A) [1]

Mark scheme for Q2.d

Correct substitution into equation:

$V = 0.2 \times 20$ [1]

Correct calculation of answer:

$V = 0.2 \times 20 = 4$ [1]

Correct units:

volts (V) [1]

Q2. a) 60 Ω

    b) 0.2 A

    c) It is the same everywhere.

    d) 4 V

    e) 8 V

    f) 12 V

Mark scheme for Q3.b and e (example for Q3.b)

Correct substitution into equation:

$I = \frac{24}{48}$ [1]

Correct calculation of answer:

$I = \frac{24}{48} = 0.5$ [1]

Correct units:

amps (A) [1]

Mark scheme for Q3.c and f (example for Q3.c)

Correct substitution into equation:

$V = 0.5 \times 20$ [1]

Correct calculation of answer:

$V = 0.5 \times 20 = 10$ [1]

Correct units:

volts (V) [1]

Q3. a) 24 V

    b) 0.5 A

    c) 10 V

    d) 24 V

    e) 1.5 A

    f) 21 V

    g) 2 A

Electrical power

Mark scheme for Q1–2 (example for Q1.a)

Correct substitution into equation:

$P = 2 \times 6$ [1]

Correct calculation of answer:

$P = 2 \times 6 = 12$ [1]

Correct units:

watts (W) [1]

Q1. a) 12 W

    b) 15 W

    c) 4.8 W

    d) 18 W

Q2. a) 125 W

    b) 31 W

    c) 630 W

    d) 25 W

Mark scheme for Q3–6 (example for Q3.a)

Correct unit conversion:

$R = 0.5 \text{ kΩ} = 500 \text{ Ω}$ [1]

Correct substitution into equation:

$I = \sqrt{\frac{500}{500}}$ [1]

Correct calculation of answer:

$I = \sqrt{\frac{500}{500}} = 1$ [1]

Correct units:

amps (A) [1]

Q3. a) 1 A

    b) 0.35 A

    c) 0.58 A

    d) 1.6 A

Q4. a) 20 V

b) 63 V

c) 13 V

d) 6.7 V

Q5. 8.7 A

Q6. 200 Ω

Mark scheme for Q7

Correct substitution into equation:

$I = \frac{5}{25}$ [1]

Correct calculation of answer:

$I = \frac{5}{25} = 0.2$ A [1]

Correct substitution into equation:

$P = 0.2 \times 5$ [1]

Correct calculation of answer:

$P = 0.2 \times 5 = 1$ [1]

Correct units:

watts (W) [1]

Q7. 1 W

Mark scheme for Q8

Correct unit conversions:

$W = 50$ kJ $= 50\ 000$ J

$t = 2$ minutes $= 120$ s [1]

Correct substitution into equation:

$P = \frac{50\ 000}{120}$ [1]

Correct calculation of answer:

$P = \frac{50\ 000}{120} = 416.7$ W [1]

Correct substitution into equation:

$I = \frac{416.7}{230}$ [1]

Correct calculation of answer:

$I = \frac{416.7}{230} = 1.8$ [1]

Correct units:

amps (A) [1]

Q8. 1.8 A

Energy transfer

Mark scheme for Q1–2 (example for Q1.a)

Correct substitution into equation:

$E = 4000 \times 6$ [1]

Correct calculation of answer:

$E = 4000 \times 6 = 24\ 000$ [1]

Correct units:

joules (J) [1]

Q1. a) 24 000 J

b) 6000 J

c) 80 000 J

d) 180 000 J

Q2. a) 9200 J

b) 120 000 J

c) 140 000 J

d) 690 J

Mark scheme for Q3–6 (example for Q3.a)

Correct unit conversion:

$E = 4.6$ MJ $= 4\ 600\ 000$ J [1]

Correct substitution into equation:

$Q = \frac{4\ 600\ 000}{230}$ [1]

Correct calculation of answer:

$Q = \frac{4\ 600\ 000}{230} = 20\ 000$ [1]

Correct units:

coulombs (C) [1]

Q3. a) 20 000 C

b) 17 000 C

c) 5200 C

d) 1000 C

Q4. a) 50 C

b) 2000 C

c) 2500 C

d) 140 C

Q5. 30 000 C

Q6. 12 V

Mark scheme for Q7

Correct substitution into equation:

$V = 0.2 \times 60$ [1]

Correct calculation of answer:

$V = 0.2 \times 60 = 12$ V [1]

Correct unit conversion:

$Q = 500$ kC $= 500\ 000$ C [1]

Correct substitution into equation:

$E = 500\ 000 \times 12$ [1]

Correct calculation of answer:

$E = 500\ 000 \times 12 = 6\ 000\ 000$ [1]

Correct units:

joules (J) [1]

Q7. 6 000 000 J

Mark scheme for Q8

Correct unit conversion:

Answers

t = 5 minutes = 300 s [1]

Correct substitution into equation:

Q = 0.5 × 300 [1]

Correct calculation of answer:

Q = 0.5 × 300 = 150 C [1]

Correct substitution into equation:

E = 150 × 230 [1]

Correct calculation of answer:

E = 150 × 230 = 35 000 [1]

Correct units:

joules (J) [1]

Q8. 35 000 J

Density

Mark scheme for Q1–2 (example for Q1.a)

Correct substitution into equation:

$\rho = \frac{800}{0.5}$ [1]

Correct calculation of answer:

$\rho = \frac{800}{0.5} = 1600$ [1]

Correct units:

kilograms per metre cubed (kg/m^3) [1]

Q1. a) 1600 kg/m^3

b) 400 kg/m^3

c) 1000 kg/m^3

d) 500 kg/m^3

Q2. a) 100 kg/m^3

b) 20 kg/m^3

c) 5 kg/m^3

d) 4 kg/m^3

Mark scheme for Q3–6 (example for Q3.a)

Correct unit conversion:

m = 100 g = 0.1 kg [1]

Correct substitution into equation:

$V = \frac{0.1}{12\,000}$ [1]

Correct calculation of answer:

$V = \frac{0.1}{12\,000} = 8.3 \times 10^{-6}$ [1]

Correct units:

metres cubed (m^3) [1]

Q3. a) 8.3 × 10^{-6} m^3

b) 6.7 × 10^{-5} m^3

c) 4.2 × 10^{-5} m^3

d) 3.3 × 10^{-6} m^3

Q4. a) 0.3 kg

b) 1.5 kg

c) 4 kg

d) 0.8 kg

Q5. 3.3 × 10^{-4} m^3

Q6. 3.1 × 10^{-12} m^3

Mark scheme for Q7

Correct unit conversion:

V = 52 cm^3 = 5.2 × 10^{-5} m^3 [1]

Correct unit conversion:

m = 800 g = 0.8 kg [1]

Correct substitution into equation:

$\rho = \frac{0.8}{5.2 \times 10^{-5}}$ [1]

Correct calculation of answer:

$\rho = \frac{0.8}{5.2 \times 10^{-5}} = 15\,000$ [1]

Correct units:

kilograms per metre cubed (kg/m^3) [1]

Correct conclusion:

No, the crown is not made of gold. [1]

Q7. 15 000 kg/m^3, not made of gold

Mark scheme for Q8

Correct unit conversion:

E_p = 784 kJ = 784 000 J [1]

Correct substitution into equation:

$m = \frac{784\,000}{9.8 \times 20}$ [1]

Correct calculation of answer:

$m = \frac{784\,000}{9.8 \times 20} = 4000$ kg [1]

Correct substitution into equation:

$\rho = \frac{4000}{2}$ [1]

Correct calculation of answer:

$\rho = \frac{4000}{2} = 2000$ [1]

Correct units:

kilograms per metre cubed (kg/m^3) [1]

Q8. 2000 kg/m^3

Specific latent heat

Mark scheme for Q1–4 (example for Q1.a)

Correct substitution into equation:

E = 0.02 × 213 000 [1]

Correct calculation of answer:

E = 0.02 × 213 000 = 4300 [1]

Correct units:

joules (J) [1]

Q1. a) 4300 J

    b) 21 000 J

    c) 17 000 J

    d) 47 000 J

Q2. a) 7000 J

    b) 2800 J

    c) 11 000 J

    d) 560 J

Q3. a) 2 kg

    b) 0.5 kg

    c) 0.33 kg

    d) 0.17 kg

Q4. a) 0.05 kg

    b) 0.25 kg

    c) 2 kg

    d) 0.4 kg

Mark scheme for Q5–7 (example for Q5)

Correct unit conversion:

E = 400 kJ = 400 000 J [1]

Correct unit conversion:

m = 2000 g = 2 kg [1]

Correct substitution into equation:

$L = \frac{400\,000}{2}$ [1]

Correct calculation of answer:

$L = \frac{400\,000}{2}$ = 200 000 J/kg [1]

Q5. 200 000 J/kg

Q6. 570 000 J/kg

Q7. 2 300 000 J/kg

Pressure in gases

Mark scheme for Q1 (example for Q1.a)

Correct substitution into equation:

constant = 200 000 × 0.02 [1]

Correct calculation of answer:

constant = 200 000 × 0.02 = 4000 [1]

Correct units:

pascal metres cubed (Pa m^3) [1]

Q1. a) 4000 Pa m^3

    b) 5000 Pa m^3

    c) 17 000 Pa m^3

    d) 6000 Pa m^3

Mark scheme for Q2–3 (example for Q2.a)

Correct substitution into equation:

constant = 120 000 × 0.04 [1]

Correct calculation of constant:

constant = 120 000 × 0.04 = 4800 Pa m^3 [1]

Correct substitution:

125 000 × V = 4800 [1]

Correct rearrangement and calculation:

$V = \frac{4800}{125\,000}$ = 0.038 [1]

Correct units:

metres cubed (m^3) [1]

Q2. a) 0.038 m^3

    b) 0.03 m^3

    c) 0.034 m^3

    d) 0.02 m^3

Q3. a) 2.0 × 10^5 Pa

    b) 2.6 × 10^4 Pa

    c) 1.3 × 10^5 Pa

    d) 6.4 × 10^4 Pa

Mark scheme for Q4–5 (example for Q4)

Correct unit conversion:

V = 200 cm^3 = 0.0002 m^3 [1]

Correct substitution into equation:

constant = 800 000 × 0.0002 [1]

Correct calculation of constant:

constant = 800 000 × 0.0002 = 160 Pa m^3 [1]

Correct substitution:

P × 0.2 = 160 [1]

Correct rearrangement and calculation:

$P = \frac{160}{0.2}$ = 800 [1]

Correct units:

pascals (Pa) [1]

Q4. 800 Pa

Q5. 380 000 Pa

Nuclear equations

Q1. a) Alpha

    b) Gamma

    c) Beta

    d) Neutron

Q2. a) Beta

    b) Neutron

c) Alpha

d) Gamma

Mark scheme for Q3–7: 1 mark for each correct mass and atomic number

Q3. a) $^{241}_{95}\text{Am} \rightarrow ^{237}_{93}\text{Np} + ^{4}_{2}\text{He}$

b) $^{222}_{88}\text{Ra} \rightarrow ^{218}_{86}\text{Rn} + ^{4}_{2}\text{He}$

c) $^{218}_{84}\text{Po} \rightarrow ^{214}_{82}\text{Pb} + ^{4}_{2}\text{He}$

d) $^{211}_{83}\text{Bi} \rightarrow ^{207}_{81}\text{Tl} + ^{4}_{2}\text{He}$

Q4. a) $^{6}_{2}\text{He} \rightarrow ^{6}_{3}\text{Li} + ^{0}_{-1}\text{e}$

b) $^{14}_{6}\text{C} \rightarrow ^{14}_{7}\text{N} + ^{0}_{-1}\text{e}$

c) $^{209}_{82}\text{Pb} \rightarrow ^{209}_{83}\text{Bi} + ^{0}_{-1}\text{e}$

d) $^{52}_{26}\text{Fe} \rightarrow ^{52}_{27}\text{Co} + ^{0}_{-1}\text{e}$

Q5. $^{238}_{92}\text{U} \rightarrow ^{234}_{90}\text{Th} + ^{4}_{2}\text{He}$

$^{234}_{90}\text{Th} \rightarrow ^{234}_{91}\text{Pa} + ^{0}_{-1}\text{e}$

Q6. $^{214}_{82}\text{Pb} \rightarrow ^{214}_{83}\text{Bi} + ^{0}_{-1}\text{e}$

$^{214}_{83}\text{Bi} \rightarrow ^{214}_{84}\text{Po} + ^{0}_{-1}\text{e}$

Q7. $^{222}_{86}\text{Rn} \rightarrow ^{218}_{84}\text{Po} + ^{4}_{2}\text{He}$

$^{218}_{84}\text{Po} \rightarrow ^{214}_{82}\text{Pb} + ^{4}_{2}\text{He}$

Half-lives

Mark scheme for Q1–3 (example for Q1.a)

Correct count rate:

$constant$ = 400 [1]

Correct units:

becquerel (Bq) [1]

Q1. a) 400 Bq

b) 200 Bq

c) 100 Bq

d) 50 Bq

Q2. a) 10 000 Bq

b) 5000 Bq

c) 2500 Bq

d) 1300 Bq

Q3. a) 17 100 years

b) 11 400 years

c) 22 800 years

d) 5700 years

Mark scheme for Q4–6 (example for Q4)

Correct number of half-lives:

2 half-lives [1]

Correct amount of time:

2400 years [1]

Q4. 2400 years

Q5. 120 s

Q6. 450 s

Mark scheme for Q7–8 (example for Q7)

Correct unit conversion:

2 minutes = 120 seconds [1]

Correct number of half-lives:

4 half-lives [1]

Correct count rate:

5 Bq [1]

Q7. 5 Bq

Q8. 45 days

Weight

Mark scheme for Q1–2 (example for Q1.a)

Correct substitution into equation:

$W = 2 \times 9.8$ [1]

Correct calculation of answer:

$W = 2 \times 9.8 = 20$ [1]

Correct units:

newtons (N) [1]

Q1. a) 20 N

b) 4.9 N

c) 3900 N

d) 150 N

Q2. a) 3.2 N

b) 0.8 N

c) 640 N

d) 24 N

Q3. Weights are lower on the Moon as there is a lower gravitational field strength.

Mark scheme for Q4–7 (example for Q4.a)

Correct unit conversion:

m = 400 g = 0.4 kg [1]

Correct substitution into equation:

$W = 0.4 \times 9.8$ [1]

Correct calculation of answer:

$W = 0.4 \times 9.8 = 3.9$ [1]

Correct units:

newtons (N) [1]

Q4. a) 3.9 N

b) 5.9 N

c) 8.8 N

d) 14 N

Q5. a) 3.7 N/kg

b) 8.8 N/kg

c) 9.8 N/kg

d) 1.6 N/kg

Q6. 0.0049 N

Q7. 1300 kg

Mark scheme for Q8.a

Correct unit conversion:

E_p = 78.4 kJ = 78 400 J   [1]

Correct substitution into equation:

$m = \frac{78\,400}{9.8 \times 40}$ [1]

Correct calculation of answer:

$m = \frac{78\,400}{9.8 \times 40}$ = 200 kg   [1]

Correct substitution into equation:

W = 200 × 9.8   [1]

Correct calculation of answer:

W = 200 × 9.8 = 2000   [1]

Correct units:

newtons (N)   [1]

Q8. a) 2000 N

Mark scheme for Q8.b

Correct substitution into equation:

$v = \sqrt{\frac{78\,400}{0.5 \times 200}}$ [1]

Correct calculation of answer:

$v = \sqrt{\frac{78\,400}{0.5 \times 200}}$ = 28   [1]

Correct units:

metres per second (m/s)   [1]

Q8. b) 28 m/s

Resultant forces

Mark scheme for Q1–4 (example for Q1.a)

Correct calculation of resultant force:

50 + 40 = 90 N   [1]

Correct direction:

Right   [1]

Q1. a) 90 N right

b) 100 N right

c) 130 N right

d) 60 N right

Q2. a) 30 000 N north

b) 25 000 N north

c) 40 000 N north

d) 80 000 N north

Q3. a) 20 N right

b) 100 N left

c) 250 N right

d) 460 N right

Q4. a) 5 × 10^6 N upwards

b) 1.5 × 10^7 N upwards

c) 2.7 × 10^7 N upwards

d) 2.9 × 10^7 N upwards

Mark scheme for Q5

Set appropriate scale and draw westerly and southerly forces.   [1]

Measure resultant force to be 8.5 N (range of 8.2–8.8 N allowed).   [1]

Give direction of force to be south-west.   [1]

Q5. 8.5 N south-west (range of 8.2–8.8 N allowed)

Work done

Mark scheme for Q1–2 (example for Q1.a)

Correct substitution into equation:

W = 120 × 50   [1]

Correct calculation of answer:

W = 120 × 50 = 6000   [1]

Correct units:

joules (J)   [1]

Q1. a) 6000 J

b) 1200 J

c) 24 000 J

d) 4800 J

Q2. a) 16 000 J

b) 4000 J

c) 240 000 J

d) 360 000 J

Mark scheme for Q3–6 (example for Q3.a)

Correct unit conversion:

F = 2 kN = 2000 N   [1]

Correct substitution into equation:

$s = \frac{80\,000}{2000}$ [1]

Correct calculation of answer:

$s = \frac{80\,000}{2000}$ = 40   [1]

Correct units:

metres (m) [1]

Q3. a) 40 m

  b) 13 m

  c) 160 m

  d) 100 m

Q4. a) 10 m

  b) 2.5 m

  c) 25 m

  d) 40 m

Q5. 60 000 N

Q6. 4 800 000 J

Mark scheme for Q7.a

Correct unit conversion:

1.2 kN = 1200 N [1]

Correct calculation of resultant force:

800 + 800 + 1200 = 2800 N [1]

Correct direction:

Right [1]

Q7. a) 2800 N right

Mark scheme for Q7.b

Correct unit conversion:

s = 1.4 km = 1400 m [1]

Correct substitution into equation:

W = 2800 × 1400 [1]

Correct calculation of answer:

W = 2800 × 1400 = 3 920 000 [1]

Correct units:

joules (J) [1]

Q7. b) 3 900 000 J

Forces and elasticity

Mark scheme for Q1–2 (example for Q1.a)

Correct substitution into equation:

F = 6 × 0.2 [1]

Correct calculation of answer:

F = 6 × 0.2 = 1.2 [1]

Correct units:

newtons (N) [1]

Q1. a) 1.2 N

  b) 0.4 N

  c) 3.6 N

  d) 1.6 N

Q2. a) 600 N

  b) 960 N

  c) 2400 N

  d) 480 N

Mark scheme for Q3–7.a (example for Q3.a)

Correct unit conversion:

F = 2 kN = 2000 N [1]

Correct substitution into equation:

$e = \dfrac{2000}{20\,000}$ [1]

Correct calculation of answer:

$e = \dfrac{2000}{20\,000} = 0.1$ [1]

Correct units:

metres (m) [1]

Q3. a) 0.1 m

  b) 0.5 m

  c) 0.13 m

  d) 0.4 m

Q4. a) 90 N/m

  b) 20 N/m

  c) 30 N/m

  d) 100 N/m

Q5. 200 N

Q6. 200 N/m

Q7. a) 1.3 N

Mark scheme for Q7.b

Correct unit conversion:

e = 5 cm = 0.05 m [1]

Correct substitution into equation:

$E_e = \dfrac{1}{2} × 25 × 0.05^2$ [1]

Correct calculation of answer:

$E_e = \dfrac{1}{2} × 25 × 0.05^2 = 0.031$ [1]

Correct units:

joules (J) [1]

Q7. b) 0.031 J

Mark scheme for Q7.c

Correct unit conversion:

m = 2 g = 0.002 kg [1]

Correct substitution into equation:

$v = \sqrt{\dfrac{0.031}{0.5 × 0.002}}$ [1]

Correct calculation of answer:

$v = \sqrt{\dfrac{0.031}{0.5 × 0.002}} = 5.6$ [1]

Correct units:

metres per second (m/s) [1]

Q7. c) 5.6 m/s

Moments

Mark scheme for Q1 (example for Q1.a)

Correct substitution into equation:

$M = 3 \times 0.8$ [1]

Correct calculation of answer:

$M = 3 \times 0.8 = 2.4$ [1]

Correct units:

newton metres (Nm) [1]

Q1. a) 2.4 Nm

    b) 6.4 Nm

    c) 8 Nm

    d) 5.6 Nm

Mark scheme for Q2 (example for Q2.a)

Correct calculation of clockwise moment:

$M = 6 \times 3 = 18$ Nm [1]

Correct calculation of anti-clockwise moment:

$M = 10 \times 2 = 20$ Nm [1]

Correct statement of whether system is balanced:

Not balanced. [1]

Q2. a) Clockwise: 18 Nm. Anti-clockwise: 20 Nm. Not balanced.

    b) Clockwise: 120 Nm. Anti-clockwise: 120 Nm. Balanced.

    c) Clockwise: 30 Nm. Anti-clockwise: 30 Nm. Balanced.

    d) Clockwise: 90 Nm. Anti-clockwise: 80 Nm. Not balanced.

Mark scheme for Q3 (example for Q3.a)

Correct calculation of first clockwise moment:

$M = 10 \times 1.5 = 15$ Nm [1]

Correct calculation of second clockwise moment:

$M = 10 \times 3 = 30$ Nm [1]

Correct total clockwise moment:

$M = 15 + 30 = 45$ Nm [1]

Correct calculation of anti-clockwise moment:

$M = 20 \times 2 = 40$ Nm [1]

Correct statement of whether system is balanced:

Not balanced. [1]

Q3. a) Clockwise: 15 + 30 = 45 Nm. Anti-clockwise: 40 Nm. Not balanced.

    b) Clockwise: 45 Nm. Anti-clockwise: 15 + 30 = 45 Nm. Balanced.

    c) Clockwise: 3 + 12 = 15 Nm. Anti-clockwise: 15 Nm. Balanced.

    d) Clockwise: 30 + 30 = 60 Nm. Anti-clockwise: 60 Nm. Balanced.

Pressure and pressure in a fluid

Mark scheme for Q1 (example for Q1.a)

Correct substitution into equation:

$p = \frac{100}{0.02}$ [1]

Correct calculation of answer:

$p = \frac{100}{0.02} = 5000$ [1]

Correct units:

pascals (Pa) [1]

Q1. a) 5000 Pa

    b) 20 000 Pa

    c) 13 000 Pa

    d) 15 000 Pa

Mark scheme for Q2–4 (example for Q2.a)

Correct unit conversion:

$h = 20$ cm $= 0.2$ m [1]

Correct substitution into equation:

$p = 0.2 \times 1020 \times 9.8$ [1]

Correct calculation of answer:

$p = 0.2 \times 1020 \times 9.8 = 2000$ [1]

Correct units:

pascals (Pa) [1]

Q2. a) 2000 Pa

    b) 5000 Pa

    c) 12 000 Pa

    d) 8500 Pa

Q3. a) 25 N

    b) 100 N

    c) 50 N

    d) 125 N

Q4. a) 4.9 m

    b) 3.9 m

    c) 1.5 m

    d) 2.1 m

Mark scheme for Q5

Correct substitution into equation:

$\rho = \frac{3000}{6}$ [1]

Correct calculation of answer:

$\rho = \frac{3000}{6} = 500$ kg/m^3 [1]

Correct unit conversion:

h = 40 cm = 0.4 m [1]

Correct substitution into equation:

p = 0.4 × 500 × 9.8 [1]

Correct calculation of answer:

p = 0.4 × 500 × 9.8 = 2000 [1]

Correct units:

pascals (Pa) [1]

Q5. 2000 Pa

Speed

Mark scheme for Q1–2 (example for Q1.a)

Correct substitution into equation:

s = 3 × 20 [1]

Correct calculation of answer:

s = 3 × 20 = 60 [1]

Correct units:

metres per second (m/s) [1]

Q1. a) 60 m

b) 150 m

c) 360 m

d) 450 m

Q2. a) 120 m

b) 300 m

c) 720 m

d) 900 m

Mark scheme for Q3–7 (example for Q3.a)

Correct unit conversion:

s = 1.2 km = 1200 m [1]

Correct substitution into equation:

$v = \frac{1200}{45}$ [1]

Correct calculation of answer:

$v = \frac{1200}{45} = 27$ [1]

Correct units:

metres per second (m/s) [1]

Q3. a) 27 m/s

b) 40 m/s

c) 24 m/s

d) 13 m/s

Q4. a) 17 m/s

b) 25 m/s

c) 6.3 m/s

d) 10 m/s

Q5. 31 m/s

Q6. 170 m/s

Q7. 5.6 m/s

Mark scheme for Q8

Correct unit conversions:

s = 20 km = 20 000 m

t = 15 minutes = 900 s [1]

Correct substitution into equation:

$v = \frac{20\ 000}{900}$ [1]

Correct calculation of answer:

$v = \frac{20\ 000}{900} = 22.2$ m/s [1]

Correct substitution into equation:

$E_k = \frac{1}{2} × 1200 × 22.2^2$ [1]

Correct calculation of answer:

$E_k = \frac{1}{2} × 1200 × 22.2^2 = 300\ 000$ [1]

Correct units:

joules (J) [1]

Q8. 300 000 J

Distance–time graphs

Q1. a) 30 m

b) 100 m

Mark scheme for Q2 (example for Q2.a)

Correct drawing of gradient triangle. [1]

Correct calculation of gradient:

$\frac{30}{10} = 3$ [1]

Correct units:

metres per second (m/s) [1]

Q2. a) 3 m/s

b) 10 m/s

Mark scheme for Q3 (example for Q3.a)

Correct drawing of gradient triangle. [1]

Correct unit conversion:

4 km = 4000 m [1]

Correct calculation of gradient:

$\frac{4000}{10} = 400$ m/s [1]

Q3. a) 400 m/s

    b) 700 m/s

Mark scheme for Q4 (example for Q4.a)

Correct drawing of tangent.

As shown in diagram.

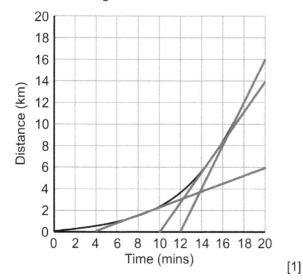

[1]

Correct drawing of gradient triangle.   [1]

Correct unit conversion:

6 km = 6000 m   [1]

Correct calculation of gradient:

$\frac{6000}{16}$ = 380 m/s (allow range of 350–410 m/s)   [1]

Q4. a) 380 m/s (allow range of 350–410 m/s)

    b) 1400 m/s (allow range of 1300–1500 m/s)

    c) 2000 m/s (allow range of 1800–2200 m/s)

Acceleration

Mark scheme for Q1 and Q2.b (example for Q1.a)

Correct substitution into equation:

$a = \frac{20}{10}$ [1]

Correct calculation of answer:

$a = \frac{20}{10} = 2$ [1]

Correct units:

metres per second squared (m/s^2)   [1]

Q1. a) 2 m/s^2

    b) 5 m/s^2

    c) 0.67 m/s^2

    d) 1.7 m/s^2

Q2. a) 4 m/s

    b) 0.5 m/s^2

Mark scheme for Q3–6 (example for Q3.a)

Correct unit conversion:

t = 3 minutes = 180 s   [1]

Correct substitution into equation:

$a = \frac{1.5}{180}$ [1]

Correct calculation of answer:

$a = \frac{1.5}{180} = 0.0083$ [1]

Correct units:

metres per second squared (m/s^2)   [1]

Q3. a) 0.0083 m/s^2

    b) 0.0042 m/s^2

    c) 0.005 m/s^2

    d) 0.025 m/s^2

Q4. a) 40 000 m/s^2

    b) 60 000 m/s^2

    c) 10 000 m/s^2

    d) 150 000 m/s^2

Q5. a) 80 000 m/s^2

    b) 16 000 m/s^2

    c) 48 000 m/s^2

    d) 320 000 m/s^2

Q6. a) 60 s

    b) 48 s

    c) 40 s

    d) 240 s

Mark scheme for Q7

Correct substitution into equation:

$\Delta v = 2 \times 2$ [1]

Correct calculation of answer:

$\Delta v = 2 \times 2 = 4$ m/s   [1]

Correct substitution into equation:

$E_k = \frac{1}{2} \times 65 \times 7^2$ [1]

Correct calculation of answer:

$E_k = \frac{1}{2} \times 65 \times 7^2 = 1600$ [1]

Correct units:

joules (J)   [1]

Q7. 1600 J

Velocity–time graphs

Mark scheme for Q1 (example for Q1.a)

Correct drawing of gradient triangle.   [1]

Correct calculation of gradient:

7 ÷ 10 = 0.7   [1]

Correct units:

metres per second squared (m/s^2) [1]

Q1. a) 0.7 m/s^2

   b) −1 m/s^2

Mark scheme for Q2–3 (example for Q2.a)

Correct working of area:

$\frac{1}{2}$ × 10 × 0.7 [1]

Correct calculation of area:

$\frac{1}{2}$ × 10 × 0.7 = 3.5 [1]

Correct units:

metres (m) [1]

Q2. a) 3.5 m

   b) 32 m

Q3. a) 9 + 42 = 51 m

   b) 5 + 20 + 10 = 35 m

Mark scheme for Q4

Correct substitution into equation:

$a = \frac{6}{3}$ [1]

Correct calculation of answer:

$a = \frac{6}{3} = 2$ [1]

Correct units:

metres per second squared (m/s^2) [1]

Q4. 2 m/s^2

Correct substitution into equation:

$2.5^2 - 0.5^2 = 2 \times a \times 2000$ [1]

Correct calculation of answer:

$a = 0.0015$ [1]

Correct units:

metres per second squared (m/s^2) [1]

Q3. a) 0.0015 m/s^2

   b) 0.001 m/s^2

   c) 0.0006 m/s^2

   d) 0.0038 m/s^2

Mark scheme for Q4

Correct substitution into equation:

$a = \frac{9}{3}$ [1]

Correct calculation of answer:

$a = \frac{9}{3} = 3$ m/s^2 [1]

Correct substitution into equation:

$9^2 - 0^2 = 2 \times 3 \times s$ [1]

Correct calculation of answer:

$s = 13.5$ [1]

Correct units:

metres (m) [1]

Q4. 14 m

Equation of motion

Mark scheme for Q1–2 (example for Q1.a)

Correct substitution into equation:

$v^2 - 0^2 = 2 \times 0.8 \times 10$ [1]

Correct calculation of answer:

$v = 4$ [1]

Correct units:

metres per second (m/s) [1]

Q1. a) 4 m/s

   b) 6.3 m/s

   c) 8 m/s

   d) 5.9 m/s

Q2. a) 35 m/s

   b) 58 m/s

   c) 40 m/s

   d) 73 m/s

Mark scheme for Q3 (example for Q3.a)

Correct unit conversion:

$s = 2$ km = 2000 m [1]

Newton's second law

Mark scheme for Q1–3 (example for Q1.a)

Correct substitution into equation:

$F = 2500 \times 2$ [1]

Correct calculation of answer:

$F = 2500 \times 2 = 5000$ [1]

Correct units:

newtons (N) [1]

Q1. a) 5000 N

   b) 4000 N

   c) 1300 N

   d) 6000 N

Q2. a) 2 m/s^2

   b) 3 m/s^2

   c) 0.25 m/s^2

   d) 8 m/s^2

Q3. a) 0.67 kg

   b) 0.5 kg

   c) 0.8 kg

d) 2 kg

Mark scheme for Q4–5 (example for Q4.a)

Correct unit conversion:

$m = 800$ g $= 0.8$ kg [1]

Correct substitution into equation:

$a = \frac{5}{0.8}$ [1]

Correct calculation of answer:

$a = \frac{5}{0.8} = 6.3$ [1]

Correct units:

metres per second squared (m/s²) [1]

Q4. a) 6.3 m/s²

    b) 10 m/s²

    c) 4.2 m/s²

    d) 7.1 m/s²

Q5. a) 1.3 m/s²

    b) 0.93 m/s²

    c) 2 m/s²

    d) 0.53 m/s²

Mark scheme for Q6

Correct substitution into equation:

$a = \frac{10}{4}$ [1]

Correct calculation of answer:

$a = \frac{10}{4} = 2.5$ m/s² [1]

Correct substitution into equation:

$F = 80 \times 2.5$ [1]

Correct calculation of answer:

$F = 80 \times 2.5 = 200$ [1]

Correct units:

newtons (N) [1]

Q6. 200 N

Stopping distance

Q1. a) 23 m

    b) 29 m

Q2. a) 23 m

    b) 30 m

Q3. a) 18 m

    b) 25 m

Mark scheme for Q4 and 6 (example for Q4.a)

Correct unit conversion:

$t = 120$ ms $= 0.12$ s [1]

Correct substitution into equation:

$s = 20 \times 0.12$ [1]

Correct calculation of answer:

$s = 20 \times 0.12 = 2.4$ [1]

Correct units:

metres (m) [1]

Q4. a) 2.4 m

    b) 3.6 m

    c) 4 m

    d) 5 m

Q5. The thinking distance increases.

Q6. a) 94 m

    b) 80 m

    c) 140 m

    d) 230 m

Mark scheme for Q7.a and Q7.b (example for Q7.a)

Correct substitution into equation:

$E_k = \frac{1}{2} \times 1100 \times 10^2$ [1]

Correct calculation of answer:

$E_k = \frac{1}{2} \times 1100 \times 10^2 = 55\,000$ [1]

Correct units:

joules (J) [1]

Q7. a) 55 000 J

    b) 220 000 J

    c) It quadruples.

    d) $W = Fs$

    e) Work done needed to brake car quadruples. [1] Therefore for a constant force, braking distance must also quadruple. [1]

Momentum

Mark scheme for Q1 (example for Q1.a)

Correct substitution into equation:

$p = 5 \times 2$ [1]

Correct calculation of answer:

$p = 5 \times 2 = 10$ [1]

Correct units:

kilograms metre per second (kg m/s) [1]

Q1. a) 10 kg m/s

    b) 16 kg m/s

    c) 12 kg m/s

    d) 9 kg m/s

Mark scheme for Q2–3 (example for Q2.a)

Correct unit conversion:

m = 56 g = 0.056 kg [1]

Correct substitution into equation:

v = 2.2 ÷ 0.056 [1]

Correct calculation of answer:

v = 2.2 ÷ 0.056 = 39 [1]

Correct units:

metres per second (m/s) [1]

Q2. a) 39 m/s

b) 43 m/s

c) 29 m/s

d) 16 m/s

Q3. a) 30 000 kg

b) 25 000 kg

c) 43 000 kg

d) 38 000 kg

Mark scheme for Q4–5 (example for Q4.a)

Correct unit conversion:

m = 1.2 g = 0.0012 kg [1]

Correct momentum calculation:

p = 0.0012 × 75 = 0.09 kg m/s [1]

Correct substitution into equation:

0.09 = 1.5 × v [1]

Correct calculation of answer:

v = 0.09 ÷ 1.5 = 0.06 [1]

Correct units:

metres per second (m/s) [1]

Q4. a) 0.06 m/s

b) 0.072 m/s

c) 0.084 m/s

d) 0.088 m/s

Q5. 5.6 m/s

Changes in momentum

Mark scheme for Q1 (example for Q1.a)

Correct substitution into equation:

$F = \frac{2200 \times 12}{0.05}$ [1]

Correct calculation of answer:

$F = \frac{2200 \times 12}{0.05}$ = 530 000 [1]

Correct units:

newtons (N) [1]

Q1. a) 530 000 N

b) 88 000 N

c) 66 000 N

d) 38 000 N

Q2. a) Force decreases.

b) Air bags spread collision over longer time [1] and reduce the force/rate of change of momentum for the person. [1]

Mark scheme for Q3 and Q5 (example for Q3.a)

Correct unit conversion:

m = 160 g = 0.16 kg [1]

Correct substitution into equation:

$t = \frac{0.16 \times 40}{640}$ [1]

Correct calculation of answer:

$t = \frac{0.16 \times 40}{640}$ = 0.01 [1]

Correct units:

seconds (s) [1]

Mark scheme for Q4 and Q6 (example for Q4.a)

Correct unit conversion:

m = 56 g = 0.056 kg [1]

Correct unit conversion:

t = 5 ms = 0.005 s [1]

Correct substitution into equation:

$\Delta v = \frac{400 \times 0.005}{0.056}$ [1]

Correct calculation of answer:

$\Delta v = \frac{400 \times 0.005}{0.056}$ = 36 [1]

Correct units:

metres per second (m/s) [1]

Q3. a) 0.01 s

b) 0.012 s

c) 0.008 s

d) 0.025 s

Q4. a) 36 m/s

b) 29 m/s

c) 43 m/s

d) 17 m/s

Q5. 0.45 s

Q6. 0.00092 s

Mark scheme for Q7

Correct unit conversion

V = 1500 cm^3 = 0.0015 m^3 [1]

Correct substitution into equation:

$m = 2000 \times 0.0015$ [1]

Correct calculation of answer:

$m = 2000 \times 0.0015 = 3$ kg [1]

Correct substitution into equation:

$F = \frac{3 \times 8}{0.002}$ [1]

Correct calculation of answer:

$F = \frac{3 \times 8}{0.002} = 12\,000$ [1]

Correct units:

newtons (N) [1]

Q7. 12 000 N

Frequency and time period

Mark scheme for Q1 (example for Q1.a)

Correct substitution into equation:

$T = \frac{1}{0.2}$ [1]

Correct calculation of answer:

$T = \frac{1}{0.2} = 5$ [1]

Correct units:

seconds (s) [1]

Q1. a) 5 s

    b) 0.02 s

    c) 0.01 s

    d) 0.25 s

Mark scheme for Q2–8 (example for Q2.a)

Correct unit conversion:

$f = 4$ kHz $= 4000$ Hz [1]

Correct substitution into equation:

$T = \frac{1}{4000}$ [1]

Correct calculation of answer:

$T = \frac{1}{4000} = 2.5 \times 10^{-4}$ [1]

Correct units:

seconds (s) [1]

Q2. a) 2.5×10^{-4} s

    b) 5×10^{-5} s

    c) 2×10^{-3} s

    d) 4×10^{-6} s

Q3. a) 10^{-6} s

    b) 8.3×10^{-8} s

    c) 2×10^{-7} s

    d) 3.3×10^{-8} s

Q4. a) 500 Hz

    b) 25 Hz

    c) 200 Hz

    d) 4 Hz

Q5. a) 5000 Hz

    b) 8300 Hz

    c) 20 000 Hz

    d) 56 000 Hz

Q6. 1.0×10^{-8} s

Q7. 3.3×10^{-12} s

Q8. 250

Wave speed equation

Mark scheme for Q1 (example for Q1.a)

Correct substitution into equation:

$v = 200 \times 0.4$ [1]

Correct calculation of answer:

$v = 200 \times 0.4 = 80$ [1]

Correct units:

metres per second (m/s) [1]

Q1. a) 80 m/s

    b) 2400 m/s

    c) 300 m/s

    d) 4000 m/s

Mark scheme for Q2–6 (example for Q2.a)

Correct unit conversion:

$f = 2$ kHz $= 2000$ Hz [1]

Correct substitution into equation:

$\lambda = \frac{340}{2000}$ [1]

Correct calculation of answer:

$\lambda = \frac{340}{2000} = 0.17$ [1]

Correct units:

metres (m) [1]

Q2. a) 0.17 m

    b) 0.57 m

    c) 0.068 m

    d) 0.028 m

Q3. a) 5000 Hz

    b) 7000 Hz

    c) 3000 Hz

    d) 2000 Hz

Q4. a) 1.5×10^{9} Hz

    b) 3.8×10^{11} Hz

    c) 6×10^{15} Hz

d) 5.6×10^{14} Hz

Q5. 20 m

Q6. 3.3×10^{-4} m

Mark scheme for Q7

Correct substitution into equation:

$f = \dfrac{1}{3.5 \times 10^{-15}}$ [1]

Correct calculation of answer:

$f = \dfrac{1}{3.5 \times 10^{-15}} = 2.86 \times 10^{14}$ Hz [1]

Correct substitution into equation:

$\lambda = \dfrac{3.0 \times 10^8}{2.86 \times 10^{14}}$ [1]

Correct calculation of answer:

$\lambda = \dfrac{3.0 \times 10^8}{2.86 \times 10^{14}} = 1.1 \times 10^{-6}$ [1]

Correct units:

metres (m) [1]

Q7. 1.1×10^{-6} m

Mark scheme for Q8

Correct unit conversion:

$s = 2$ km = 2000 m [1]

Correct substitution into equation:

$v = \dfrac{2000}{40}$ [1]

Correct calculation of answer:

$v = \dfrac{2000}{40} = 50$ m/s [1]

Correct substitution into equation:

$\lambda = \dfrac{50}{200}$ [1]

Correct calculation of answer:

$\lambda = \dfrac{50}{200} = 0.25$ [1]

Correct units:

metres (m) [1]

Q8. 0.25 m

Magnification

Mark scheme for Q1–2 (example for Q1.a)

Correct substitution into equation:

$magnification = \dfrac{0.06}{0.003}$ [1]

Correct calculation of answer:

$magnification = \dfrac{0.06}{0.003} = 20$ [1]

Q1. a) 20

b) 50

c) 15

d) 30

Q2. a) 20

b) 8

c) 5

d) 3.3

Mark scheme for Q3–8 (example for Q3.a)

Correct unit conversion:

$object\ height$ = 40 µm = 4×10^{-5} m [1]

Correct substitution into equation:

$magnification = \dfrac{0.002}{4 \times 10^{-5}}$ [1]

Correct calculation of answer:

$magnification = \dfrac{0.002}{4 \times 10^{-5}} = 50$ [1]

Q3. a) 50

b) 40

c) 29

d) 200

Q4. a) 0.018 m

b) 0.03 m

c) 0.048 m

d) 0.084 m

Q5. a) 0.04 m

b) 0.17 m

c) 0.13 m

d) 0.05 m

Q6. 40

Q7. 2000

Q8. 140 000

Force on a current carrying wire

Mark scheme for Q1–2 (example for Q1.a)

Correct substitution into equation:

$F = 0.2 \times 0.5 \times 0.1$ [1]

Correct calculation of answer:

$F = 0.2 \times 0.5 \times 0.1 = 0.01$ [1]

Correct units:

newtons (N) [1]

Q1. a) 0.01 N

b) 0.0025 N

c) 0.0045 N

d) 0.006 N

Q2. a) 0.03 N

b) 0.012 N

c) 0.0024 N

d) 0.078 N

Mark scheme for Q3–6 (example for Q3.a)

Correct unit conversion:

l = 15 cm = 0.15 m [1]

Correct substitution into equation:

$I = \dfrac{0.08}{0.05 \times 0.15}$ [1]

Correct calculation of answer:

$I = \dfrac{0.08}{0.05 \times 0.15} = 11$ [1]

Correct units:

amps (A) [1]

Q3. a) 11 A

b) 8 A

c) 3.2 A

d) 40 A

Q4. a) 0.0012 T

b) 0.02 T

c) 0.4 T

d) 0.015 T

Q5. 4.2 m

Q6. 7.5 T

Mark scheme for Q7

Correct substitution into equation:

$F = 8 \times 10^{-2} \times 4 \times 0.2$ [1]

Correct calculation of answer:

$F = 8 \times 10^{-2} \times 4 \times 0.2 = 0.064$ N [1]

Correct substitution into equation:

$m = \dfrac{0.064}{9.8}$ [1]

Correct calculation of answer:

$m = \dfrac{0.064}{9.8} = 0.0065$ [1]

Correct units:

kilograms (kg) [1]

Q7. 0.0065 kg

Mark scheme for Q8

Correct unit conversion:

R = 2.4 kΩ = 2400 Ω [1]

Correct substitution into equation:

$I = \dfrac{12}{2400}$ [1]

Correct calculation of answer:

$I = \dfrac{12}{2400} = 0.005$ A [1]

Correct substitution into equation:

$B = \dfrac{0.01}{0.005 \times 0.04}$ [1]

Correct calculation of answer:

$B = \dfrac{0.01}{0.005 \times 0.04} = 50$ [1]

Correct units:

tesla (T) [1]

Q8. 50 T

Transformers

Mark scheme for Q1–2 (example for Q1.a)

Correct substitution into equation:

$120 \times 0.5 = 6 \times I_p$ [1]

Correct calculation of answer:

$I_p = \dfrac{120 \times 0.5}{6} = 10$ [1]

Correct units:

amps (A) [1]

Q1. a) 10 A

b) 40 A

c) 60 A

d) 1.5 A

Q2. a) 0.15 V

b) 10 V

c) 4 V

d) 0.7 V

Mark scheme for Q3–4 (example for Q3.a)

Correct substitution into equation:

$\dfrac{3}{480} = \dfrac{100}{n_s}$ [1]

Correct calculation of answer:

n_s = 16 000 [1]

Q3. a) 16 000

b) 3200

c) 6400

d) 24 000

Q4. a) 50 000

b) 200 000

c) 400 000

d) 250 000

Mark scheme for Q5 (example for Q5.a)

Correct unit conversion:

420 kV = 420 000 V [1]

Correct substitution into equation:

$\dfrac{420\,000}{230} = \dfrac{n_p}{4600}$ [1]

Correct calculation of answer:

N_p = 8 400 000 [1]

Q5. a) 8 400 000

b) 3000 W